THE
SON OF ROYAL LANGBRITH

THE SON OF
ROYAL LANGBRITH

A NOVEL

BY

W. D. HOWELLS

AUTHOR OF
"QUESTIONABLE SHAPES" "LETTERS HOME"
"LITERARY FRIENDS AND ACQUAINTANCE"
"THEIR SILVER WEDDING JOURNEY"
ETC. ETC.

NEW YORK AND LONDON

HARPER & BROTHERS PUBLISHERS

1904

THE
SON OF ROYAL LANGBRITH

I

"WE'RE neither of us young people, I know, and I can very well believe that you had not thought of marrying again. I can account for your surprise at my offer, even your disgust—" Dr. Anther hesitated.

"Oh no!" Mrs. Langbrith protested.

"But I can't see why it should be 'terrible,' as you call it. If you had asked me simply to take 'no' for an answer, I could have taken it. Or taken it better."

He looked at her with a wounded air, and she said, "I didn't mean 'terrible' in that way. I was only thinking of it for myself, or not so much myself as — some one." She glanced at him, where, tenderly indignant with her, he stood by the window, quite across the room, and she seemed to wish to say more, but let her eyes drop without saying more.

He was silent, too, for a time which he allowed to

prolong itself in the apparent expectation that she would break their silence. But he had to speak first. "I don't like mysteries. I can forget—or ignore—any sense of 'terrible' you had in mind, if you will tell me one thing. Do you ask me *now* to take simply 'no' for an answer?"

"Oh no!" The words were as if surprised from her, and she made with her catching breath as if she would have caught them back.

He came quickly across the room to her. "What is it, Amelia?"

"I can't tell you," she shuddered out, and she recoiled, pulling herself up, as if she wished to escape but felt an impenetrable hinderance at her back. In the action, she showed taller than she was, and more girlishly slender. At forty, after her wifehood of three years and her widowhood of nineteen years, the inextinguishable innocence of girlhood, which keeps itself through all the experiences of a good woman's life, was pathetic in her appealing eyes; and the mourning, subdued to the paler shades of purple, which she permanently wore, would have made a stranger think of an orphan rather than a widow in her presence.

Anther's burly frame arrested itself at her recoil. His florid face, clean shaven at a time when nearly all men wore beards, was roughed to a sort of community of tint with his brown overcoat by the weather of many winters' and summers' driving in his country practice. His iron-gray hair, worn longer than the fashion was in towns, fell down his temples and neck from under his soft hat. He

2

had on his driving-gloves, and he had his whip in one hand. He had followed Mrs. Langbrith indoors in that figure from the gate, where his unkempt old horse stood with his mud-spattered buggy, to pursue the question which she tried rather than wished to shun, and he did not know that he had not uncovered. At the pathos in her eyes and in her cheeks, which had the vertical hollows showing oftener in youth than in later life, the harshness of gathered will went out of his face. "I know what you mean," he said; and at his words the tears began to drip down her face without the movement of any muscle in it, as if a habit of self-control which enabled her to command the inward effects of emotion had not been able to extend itself to its displays. "Poor thing!" he pitied her. "Must you always have a tyrant?"

"He isn't a tyrant," she said.

"Oh yes, he is! I know the type. I dare say he doesn't hit you, but he terrorizes you."

Mrs. Langbrith did not speak. In her reticence, even her tears stopped.

"You *tempt* him to bully you. Lord bless me, you tempt *me!* But I won't; no, I won't. Amelia, why, in Heaven's name, should he object? He has his own interests, quite apart from yours; his own world, which you couldn't enter if he would let you. A fellow in his junior year at college is as remote from his mother in everything as if he were in another planet!"

"We write to each other every Sunday," she urged, diffidently.

"I have no doubt you try to keep along with him; that's your nature; but I know that he cowed you before he left home, and I know that he cows you still. I could read your correspondence—the spirit —without seeing it. He isn't to blame. You've let him walk over you till he thinks there is no other path to manhood. Remember, I don't say Jim is a bad fellow. He is a very good fellow—considering."

The doctor went to the window and stooped to look out at his horse, which remained as he had left it, only more patiently sunken in a permanence expressed by the collapsing of its hind quarters into a comfortable droop, and a dreamy dejection of its large head. In the mean time, Mrs. Langbrith had sat down in a chair which she seemed to think had offered itself to her, and when the doctor came back he asked, "May I sit down?"

"Why, of course! I'm ashamed—"

"No, no! Don't say that! Don't say anything like that!"

In the act of sitting down, he realized that he had his hat on. He took it off and put it on the floor near his feet, where it toppled into a soft heap. His hair had partly lifted with it, and its disorder on his crown somewhat concealed its thinness. "I want to talk this matter over sensibly. We are not two young things that we need be scared at our own feelings, or each other's. I suppose I may say we both knew it was coming to something like this?"

She might not have let him say so for her, but in her silence he went on to say so for himself.

"*I* knew it was coming, anyway, and I've known

4

it for a good while. I have liked you ever since I came to Saxmills"—she trembled and colored a little—"but I wouldn't be saying what I am saying to you, if I had cared for you before Langbrith died as I care for you now. That would be, to my thinking, rather loathsome. I should despise myself for it; I should despise you; I couldn't help it. But we are both fairly outside of that. I didn't begin to realize how it was with me till about a year ago, and I don't suppose that you—"

Mrs. Langbrith shifted her position slightly, but he did not notice, and he began again.

"So I feel that I can offer you a clean hand. I'm six years older than you are, which makes it just about right; and I'm not so poor that I need seem to be after your—thirds. I've got a good practice, and I don't intend to take life so hard hereafter. I could give you as pleasant a home—"

"Oh, I couldn't think of leaving this!" she broke out, helplessly.

Anther allowed himself to smile. "Well, well, there's no hurry. But if Jim marries—"

"I should live with him."

"I'm imagining that you would live with me."

"You mustn't."

"I'm merely imagining; I'm not trying to commit you to anything, or to overrule you at all. My idea is, that there's been enough of that in your life. I want you to overrule me, and if you don't fancy settling down immediately, and would like a year or two in Europe first, I could freshen my science up in Vienna or Paris, and come back all the better

5

prepared to keep on in my practice here; or I could give up my practice altogether."

"You oughtn't to do that."

"No, I oughtn't. But all this is neither here nor there, till the great point is settled. *Do* you think any one could care for me as I care for you?"

"Why, of course, Dr. Anther?"

"Do *you* care for me—that way—now?" He seemed to expect evasion or hesitation, even such elusion as might have expressed itself in material escape from him, and, unconsciously, he hitched his chair forward as if to hem her in.

It was a needless precaution. She answered instantly, "You *know* I do."

"Amelia," Anther asked solemnly, without changing his posture or the slant of his face in its lift towards hers, "have I put any pressure on you to say this? Do you say it as freely as if I hadn't asked you?"

The absurdity of the question did not appear to either of them. She answered, "I say it as freely as if it had never been asked. I would have said it years ago. I have always liked you—that way. Or ever since—"

He leaned back in his chair and pushed his hands forward on the arms. "Then—then—" he began, bewilderedly, and she said:

"But—"

"Ah!" he broke out, "I know what that 'but' means. Why need there be any such 'but'? Do you think he dislikes me?"

"No, he likes you; he respects you. He says you

are a physician who would be famous in a large place. He—"

Anther put the rest aside with his hand. "Then he would object to any one? Is that it?"

"Yes," said Mrs. Langbrith, with a drop to specific despair from her general hopelessness.

"I don't recognize his right," Anther said sharply, "unless he is ready to promise that he will never leave you to be pushed aside; turned from a mother into a mother-in-law. I don't recognize his right. Why does he assume such a right?"

"Out of reverence for his father's memory."

II

ONE cannot look on a widow who has long survived her husband without a curiosity not easily put into terms. The curiosity is intensified and the difficulty enhanced if there are children to testify of a relation which, in the absence of the dead, has no other witness. The man has passed out of the woman's life as absolutely as if he had never been there; it is conceivable that she herself does not always think of her children as also his. Yet they are his children, and there must be times when he holds her in mortmain through them, when he is still her husband, still her lord and master. But how much, otherwise, does she keep of that intimate history of emotions, experiences, so manifold, so recondite? Is he as utterly gone, to her sense, as to all others? Or is he in some sort there still, in her ear, in her eye, in her touch? Was it for the nothing which it now seems that they were associated in the most tremendous of the human dramas, the drama that allies human nature with the creative, the divine and the immortal, on one side, the bestial and the perishable on the other? Does oblivion pass equally over the tremendous and the trivial and blur them alike?

Anther looked at Mrs. Langbrith in a whirl of

question: question of himself as well as of her. By virtue of his privity to her past, he was in a sort of authority over her; but it may have been because of his knowledge that he almost humbly forbore to use his authority.

"Amelia," he entreated her, "have you brought him up in a superstition of his father?"

"Oh no!" She had the effect of hurrying to answer him. "Oh, never!"

"I am glad of that, anyway. But if you have let him grow up in ignorance—"

"How could I help that?"

"You couldn't! *He* made himself solid, there. But the boy's reverence for his father's memory is sacrilege—"

"I know," she tremulously consented; and in her admission there was no feint of sparing the dead, of defending the name she bore, or the man whose son she had borne. She must have gone all over that ground long ago, and abandoned it. "It ought to have come out," she even added.

"Yes, but it never can come out now, while any of his victims live," Anther helplessly raged. "I'm willing to help keep it covered up in his grave myself, because you're one of them. If poor Hawberk had only taken to drink instead of opium!"

"Yes," she again consented, with no more apparent feeling for the memory imperilled by the conjecture than if she had been nowise concerned in it.

"But you must, Amelia—I hate to blame you; I know how true you are—you must have let the boy think—"

9

"As a child, he used to ask me, but not much; and what would you have had I should answer him?"

"Of course, of course! You couldn't."

"I used to wonder *if* I could. Once, when he was little, he put his finger on this"—she put her own finger on a scar over her left eye—"and asked me what made it. I almost told him."

Anther groaned and twisted in his chair. "The child always spoke of him," she went on passionlessly, "as being in heaven. I found out, one night, when I was saying his prayers with him, that—you know how children get things mixed up in their thoughts—he supposed Mr. Langbrith was the father in heaven he was praying to."

"Gracious powers!" Anther broke out.

"I suppose," she concluded, with a faint sigh, "though it's no comfort, that there are dark corners in other houses."

"Plenty," Anther grimly answered, from the physician's knowledge. "But not many as dark as in yours, Amelia."

"No," she passively consented once more. "As he grew up," she resumed the thread of their talk, without prompting, "he seemed less and less curious about it; and I let it go. I suppose I wanted to escape from it, to forget it."

"I don't blame you."

"But, doctor," she pleaded with him for the extenuation which she could not, perhaps, find in herself, "I never did teach him by any word or act—unless not saying anything was doing it—that his

10

father was the sort of man he thinks he was. I
should have been afraid that Mr. Langbrith himself would not have liked that. It would have been
a fraud upon the child."

"I don't think Langbrith would have objected to
it on that ground," Anther bitterly suggested.

"No, perhaps not. But between Mr. Langbrith
and me there were no concealments, and I felt that
he would not have wished me to impose upon the
child expressly."

"Oh, he preferred the tacit deceit, if it would serve
his purpose. I'll allow that. And in this case it
seems to have done it."

"Do you mean," she meekly asked, "that I have
deceived James?"

"No," said Anther, with a blurt of joyless laughter. "But if such a thing were possible, if it were
not too sickeningly near some wretched superstition
that doesn't believe in itself, I should say that his
father deceived him through you, that he diaboli-
cally acted through your love, and did the evil
which we have got to face now. Amelia!" he star-
tled her with the resolution expressed in his utterance of her name, "you say the boy will object to
my marrying you. Do you object to my telling
him?"

"Telling him?"

"Just what his father was!"

"Oh, you mustn't! It would make him hate
you."

"What difference?"

"I couldn't let him hate you. I couldn't bear

11

that." The involuntary tears, kept back in the abstracter passages of their talk, filled her eyes again, and trembled above her cheeks.

"If necessary, he has got to know," Anther went on, obdurately. "I won't give you up on a mere apprehension of his opposition."

"Oh, *do* give me up!" she implored. "It would be the best way."

"It would be the worst. I have a right to you, and if you care for me, as you say—"

"I do!"

"Then, Heaven help us, you have right to me. You have a right to freedom, to peace, to rest, to security; and you are going to have it. Now, will you let it come to the question without his having the grounds of a fair judgment, or shall we tell him what he ought to know, and then do what we ought to do: marry, and let me look after you as long as I live?"

She hesitated, and then said, with a sort of furtive evasion of the point: "There is something that I ought to tell you. You said that you would despise yourself if you had cared for me in Mr. Langbrith's lifetime." She always spoke of her husband, dead, as she had always addressed him, living, in the tradition of her great juniority, and in a convention of what was once polite form from wives to husbands, not to be dropped in the most solemn, the most intimate, moments.

Anther found nothing grotesque in it, and, therefore, nothing peculiarly pathetic. "Well?" he asked, impatiently trying for patience.

"Well, I know that I *cared* for you then. I couldn't help it. Now you despise me, and that ends it."

Anther rubbed his hand over his face; then he said, "I don't believe you, Amelia."

"I did," she persisted.

"Well, then, it was all right. You *couldn't* have had a wrong thought or feeling, and the theory may go. After all, I was applying the principle in my own case, and trying to equal myself with you. If you choose to equal yourself with me by saying this, I must let you; but it makes no difference. You cared for me because I stood your helper when there was no other possible, and that was right. Now, shall we tell Jim, or not?"

She looked desperately round, as if she might escape the question by escaping from the room. As all the doors were shut, she seemed to abandon the notion of flight, and said, with a deep sigh, "I must see him first."

Anther caught up his hat and put it on, and went out without any form of leave-taking. When the outer door had closed upon him, she stole to the window, and, standing back far enough not to be seen, watched him heavily tramping down the brick walk, with its borders of box, to the white gate-posts, each under its elm, budded against a sky threatening rain, and trailing its pendulous spray in the wind. He jounced into his buggy, and drew the reins over his horse, which had been standing unhitched, and drove away. She turned from the window.

III

EASTER came late that year, and the jonquils were
there before it, even in the Mid-New England latitude of Saxmills, when James Langbrith brought
his friend Falk home with him for the brief vacation.
The two fellows had a great time, as they said to
each other, among the village girls; and perhaps
Langbrith evinced his local superiority more ap-
preciably by his patronage of them than by the
colonial nobleness of the family mansion, squarely
fronting the main village street, with gardened
grounds behind dropping to the river. He did not
dispense his patronage in all cases without having
his hand somewhat clawed by the recipients, but
still he dispensed it; and, though Falk laughed when
Langbrith was scratched, still Langbrith felt that
he was more than holding his own, and he made up
for any defeat he met outside by the unquestioned
supremacy he bore within his mother's house. Her
shyness, out of keeping with her age and stature,
invited the sovereign command which Langbrith
found it impossible to refuse, though he tempered
his tyranny with words and shows of affection well
calculated to convince his friend of the perfect intelligence which existed between his mother and
himself. When he thought of it, he gave her his arm

14

in going out to dinner; and, when he forgot, he tried to make up by pushing her chair under her before she sat down. He was careful at table to have the conversation first pay its respects to some supposed interest of hers, and to return to that if it strayed afterwards, and include her. He conspicuously kissed her every morning when he came down to breakfast, and he kissed her at night when she would have escaped to bed without the rite.

It was Falk's own fault if he did not conceive from Langbrith's tenderness the ideal of what a good son should be in all points. But, as the Western growth of a German stock transplanted a generation before, he may not have been qualified to imagine the whole perfection of Langbrith's behavior from the examples shown him. His social conditions in the past may even have been such that the ceremonial he witnessed did not impress him pleasantly; but, if so, he made no sign of displeasure. He held his peace, and beyond grinning at Langbrith's shoulders, as he followed him out to the dining-room, he did not go. He seemed to have made up his mind that, without great loss of self-respect, he could suffer himself to be used in illustration of Langbrith's large-mindedness with other people whom Langbrith wished to impress. At any rate, it had been a choice between spending the Easter holiday at Cambridge, or coming home with Langbrith; and he was not sorry that he had come. He was getting as much good out of the visit as Langbrith.

One night, when Mrs. Langbrith came timidly into

15

the library to tell the two young men that dinner
was ready—she had shifted the dinner-hour, at her
son's wish, from one o'clock to seven—Langbrith
turned from the shelf where he had been looking
into various books with his friend, and said to his
mother, in giving her his arm: "I can't understand
why my father didn't have a book-plate, unless it
was to leave me the pleasure of getting one up in
good shape. I want you to design it for me, will you,
Falk?" he asked over his shoulder. Without waiting for the answer, he went on, instructively, to his
mother: "You know the name was originally Nor-
man."

"I didn't know that," she said, with a gentle self-
inculpation.

"Yes," her son explained. "I've been looking it
up. It was Longuehaleine, and they translated it
after they came to England into Longbreath, or
Langbrith, as we have it. I believe I prefer our
final form. It's splendidly suggestive for a book-
plate, don't you think, Falk?" By this time he
was pushing his mother's chair under her, and talking over her head to his friend. "A boat, with a
full sail, and a cherub's head blowing a strong gale
into it: something like that."

"They might think the name was Longboat, then,"
said Falk.

Mrs. Langbrith started.

"Oh, Falk has to have his joke," her son explained, tolerantly, as he took his place; "nobody
minds Falk. Mother, I wish you would give a dinner for him. Why not? And we could have a

dance afterwards. The old parlors would lend themselves to it handsomely. What do you say, Falk?"

"Is it for me to say I will be your honored guest?"

"Well, we'll drop that part. We won't feature you, if you prefer not. Honestly, though, I've been thinking of a dinner, mother."

Langbrith had now taken his place, and was poising the carving knife and fork over the roast turkey, which symbolized in his mother's simple tradition the extreme of formal hospitality. She wore her purple silk in honor of it, and it was what chiefly sustained her in the presence of the young men's evening dress. This was too much for her, perhaps, but not too much for the turkey. The notion of the proposed dinner, however, was something, as she conceived it, beyond the turkey's support. Without knowing just what her son meant, she cloudily imagined the dinner of his suggestion to be a banquet quite unprecedented in Saxmills society. Dinners there, except in a very few houses, were family dinners, year out and year in. They were sometimes extended to include outlying kindred, cousins and aunts and uncles who chanced to be in town or came on a visit. Very rarely, a dinner was made for some distinguished stranger: a speaker, who was going to address a political rally in the afternoon, or a lecturer, who was to be heard in the evening at the town-hall, or the clerical supply in the person of one minister or another who came to be tried for the vacant pulpit of one of the churches. Then, a few principal citizens with their wives were asked, the ministers of the other churches, the bank president, some leading merchant, the magnates of the law or medicine. The dinner was at one o'clock, and the young people were rigidly excluded. They were fed either before or after it, or farmed out among the neighboring houses till the guests were gone. Ordinarily, guests were asked to tea, which was high, with stewed chicken, hot bread, made dishes and several kinds of preserves and sweet pickles, with many sorts of cake. The last was the criterion of tasteful and lavish hospitality.

Clearly, it was nothing of all this that Mrs. Langbrith's son had in mind. After his first year in college, when he had been so homesick that everything seemed perfect under his mother's roof in his vacation visits, he began to bring fellows with him. Then he began to make changes. The dinner-hour was advanced from mid-day to evening, and he and his friends dressed for it. He had still to carve, for the dinner in courses was really unmanageable and unimaginable in his mother's house-keeping; but he professed a baronial preference for carving, and he fancied an old-fashioned, old-family effect from it. The service was such as the frightened inexperience of the elderly Irish second-girl could render; under Langbrith's threatening eye, she succeeded in offering the dishes at the left hand, and, though she stood a good way off and rather pushed them at the guests, the thing somehow was done. At least, the covered dishes were no longer set on the table, as they used always to be.

Mrs. Langbrith had witnessed the changes with

18

trepidation but absolute acquiescence even at the first, and finally with the submission in which her son held her in everything. In the afternoon, when he and his friend, whoever it might be, put on their top-hats and top-coats and went out to call on the village girls, who did not know enough of the world to offer them tea, she spent the interval before dinner in arranging for the meal with the faithful, faded Norah. After dinner, when the young men again put on their top-hats and top-coats to call again upon the village girls, whom they had impressed with the correctness of afternoon calls, and to whom they now relented in compliance with the village custom of evening calls, Mrs. Langbrith debated with Norah the success of the dinner, studied its errors, and joined her in vows for their avoidance.

IV

THE event which confronted Mrs. Langbrith in her son's words, as he sat behind the turkey and plunged the carving-fork into its steaming and streaming breast, was so far beyond the scope of her widened knowledge that she mutely waited for him to declare it.

"People," he went on, "have been so nice to Falk and me, that I think we ought to make some return. I put the duty side first, because I know you'll like that, mother, and it will help to reconcile you to the fun of it. Falk is such a pagan that he can't understand, but it will be for his good, all the same. My notion is to have a good, big dinner—twelve or fourteen at table, and then a lot in afterwards, with supper about midnight. What do you say, mother? Don't mind Falk, if you don't agree quite."

"There *is* no Falk, Mrs. Langbrith," the young fellow said, with an intelligence which comforted her and emboldened her against her son.

"I don't see—" she began, and then she stopped.

"That's right!" her son encouraged her.

"James," she said, desperately, "I wouldn't know how to do it."

"I don't *want* you to do it." He laughed exultantly. "I propose to do it myself. I will have the

20

whole thing sent up from Boston." Between her gasps, he went on: "All I have got to do is to write an order to White, the caterer, with particulars of quantity and quality, the date and the hour, and it comes on the appointed train with three men in plain clothes; two reappear in lustrous dress-suits at dinner and supper, and serve the things the other has cooked at our range. I press the button, White does the rest. He brings china, cutlery, linen— everything. All *you* have to do is to hide Jerry in the barn and keep Norah up-stairs to show the ladies into the back chamber to take off their things. You can put our own cook under the sink. You'll be astonished at the ease of the whole thing."

"Yes," Mrs. Langbrith said, "it will be easy, but—"

"But would it be right?" her son tenderly mocked. "What did I tell you?" he asked towards his friend. "In New England, the notion of ease conveys the sense of culpability. My mother is afraid she would have a bad conscience. If she took all the work and worry on herself, she would feel that she was paying the penalty of her pleasure beforehand; if she didn't, she would know that she must pay for it afterwards. Isn't that so, mother? But now you leave it to me, you dear old thing." Langbrith ran round the table and kissed her on top of the head, and made her blush like a girl, as he patted her shoulder. "Just imagine I was master, and you couldn't help yourself." He went back to his place. "What was the largest dinner you ever had in the old time?"

She hesitated, as if for his meaning. "Mr. Lang-

21

brith once entertained a company of six gentlemen, who came up here and talked of locating some cotton-mills. We called it 'supper.'"

"I can imagine them. Can't you, Falk? The moneyed man to supply the funds, the lawyer to draw up the papers, the civil engineer to survey the property. Very solemn, and a little pompous, but secretly ready for a burst if the opportunity offered under the right auspices; something like an outing of city officials."

"They were very pleasant gentlemen," Mrs. Langbrith interposed, as from her conscience.

"Oh, I dare say they were when they had tasted my father's madeira. But about our dinner now? I don't think we'd better have more than twelve, and I should want them equally divided between youth and age."

Mrs. Langbrith looked at him as if she did not quite understand him, and he said:

"Have Jessamy Colebridge and Hope Hawberk and Susie Johns and Bob Matthewson—he's a good fellow—and make out the half-dozen with Falk and me; we're both good fellows. Then, on your side of the line, yourself first of all, mother, and the rector and his wife, and Judge and Mrs. Garley, and—who else? Oh, Dr. Anther, of course! I want Falk to meet the doctor—the dearest and quaintest old type in the world. I don't know why he hasn't been in to see us, mother. Has he been here lately?"

"He was here a day or two before you came," Mrs. Langbrith answered, with her eyes down.

"Perhaps he has been waiting for me to call. Well, what do you think of my dinner-party?"

"It seems very nice," Mrs. Langbrith sighed.

"And haven't you any preferences? Nobody you want to turn down?"

"It will be a good deal of a surprise for Saxmills," she suffered herself to say.

"I flatter myself it will. I have been telling Falk that the mixed assembly of old and young is unknown in Saxmills."

Falk had not troubled himself to take part in the discussion, if it was that, but had given himself to the turkey and the cranberry sauce, with the mashed potatoes and the stewed squash, which Mrs. Langbrith had very good. Her son had obliged her to provide claret, which Falk now drank out of an abnormal glass with a stout stem and pimpled cup, hitherto dedicated to currant wine, before saying: "It astonished me less than if I had been used to something different all my life. You ought to have tried the other thing on me."

"Well, I only supposed from the smartness of the people in your *Caricature* pictures that you had always lived in a whirl of fashion."

"That shows how little you know of fashion," said Falk, and Langbrith laughed with the difficult joy of a man who owns a hit.

Mrs. Langbrith glanced from one to the other; from her son, with his long, distinguished face (he had decided that it was colonial), to the dark, aquiline type of Falk, with his black hair, his upward-pointed mustache, his pouted lips, and his prominent, floating, brown eyes. In her abeyance, she was scared at the bold person who was not afraid of her son.

"Well," said Langbrith, "I shall have to find some one to illustrate my *vers de société* who knows enough of the world for both."

"You couldn't!" Falk insinuated.

Mrs. Langbrith did not quite catch the point, but her son laughed again. "No one ever distances *you*, Falk!"

He discussed the arrangement of the affair with his mother. At the end, as she rose, obedient to his sign, and he came round to give her his arm, he said: "After all, perhaps, it wouldn't be well to strike too hard a blow. If you think you can get it up by Saturday night, mother, we'll drop the notion of having White. Make it tea, with turkey at one end of the table and chicken pie at the other, and all the sweet pickles and preserves and kinds of cake you can get together; coffee straight through, and a glass of the old Langbrith madeira to top off with."

V

Mrs. Langbrith went into the library with her son and his friend by the folding doors from the dining-room, but only to go out of the door which opened into the hall, and escape by that route to the kitchen for an immediate conference with the cook.

The young men dropped into deep leather chairs at opposite corners of the fireplace, after lighting their cigars. Probably, the comfort of his seat suggested Langbrith's reflection: "It is a shame I never knew my father. We should have had so much in common. I couldn't imagine anything more adapted to the human back than these chairs."

"His taste?" Falk asked, between whiffs.

"Everything in the house is his taste. I don't believe my mother has changed a thing. He must have been a strong personality." Langbrith followed his friend's eye in its lift towards his father's portrait over the mantel.

"I should think so," Falk assented.

"Those old New England faces," Langbrith continued, meditatively, "have a great charm. From a child, that face of my father's fascinated me. As I got on, and began to be interested in my environment, I read into it all I had read out of Hawthorne about the Puritan type. I put the grim old chaps out of *The Scarlet Letter* and *The House of the Seven Gables* and the *Twice-Told Tales* into it, and interpreted my father by them. But, really, I knew very little about him. My mother's bereavement seemed to have sealed her lips, and I preferred dreaming to asking. A kid is queer! Once or twice when I did ask, she evaded answering; that was after I was old enough to understand, and I didn't press my questions. He was much older than she; twenty years, I believe. He couldn't have been a Puritan in his creed; he was a Unitarian, as far as church-going went, and I believe my mother is a Unitarian yet; but she goes to the Episcopal Church, which makes itself a home for everybody, and she likes the rector. You'll like him, too, Falk."

"He won't talk theology to me, I suppose," Falk grumbled.

"He'll talk athletics with you. The good thing about a man of his church is that he's usually a man of the world, too. He's an Enderby, you know."

"I shouldn't be much the wiser, if I did," Falk said.

"I wouldn't work that pose so hard, Falk. You can't get even with the Enderbys by ignoring them; and you can't pretend it's meekness that makes you profess ignorance. The only thing I don't like about you is your peasant pride."

"I still have hopes of winning your whole heart then. I'll study your peasant humility."

Langbrith made as if he had not noticed the point. He rose and moved restively about the room, and then came back to his chair again, and

26

said, as if he had really been thinking of something else: "If I should decide to take up dramatic literature, I believe I'll go to Paris to continue my studies, and perhaps we'll keep on there together. I wish we could! Can't you manage it, somehow? Those things of yours in *Caricature* have attracted attention; and if *Life* has asked you to send something, why couldn't you get a lot of orders, and go out with me?"

"Gentle dreamer!" Falk murmured.

"No, but why not, really?"

"Because a lot of orders are not to be got for the asking, and I'm a bad hand at asking. I think my cheek is good for applying to a New York paper for a chance to do scenes in court, and hurry-pictures of fires, and the persons in a vivid accident; but I don't think it would hold out to invite *Harper's* or *Scribner's* to have me do high-class studies abroad for them. I may be a fool, but I am not that kind of fool. Unless," Falk hastened to anticipate, "I'm *all* kinds."

Langbrith was apparently not watching for the chance snatched from him. "Well, I think you could do it, somehow," he insisted. "I'm going to Paris for my post-graduate business, and I've set my heart on having you with me. I wonder," he mused aloud, "why I like you so much, Falk?"

"I couldn't say," Falk returned, without apparent interest in the mystery.

"You're always saying nasty things to me," Langbrith pursued. "You take every chance to give me a dig."

"It's all that keeps you in bounds."

"No—"

"Yes, it is; your arrogance would naturally splay all over the place. But just at present, you're in the melting mood that saps everybody's manhood towards the end of the senior year. If I didn't watch myself, I should feel a tenderness for *you* at times."

"Would you, really, Falk?" Langbrith appeared touched, and interested.

"I shall never know, for I don't mean to be taken off my guard."

"What a delightful fellow you are, Falk!"

"Do you think so? I should suppose you were a woman."

"Oh, it isn't the women alone that love you, old man. I love you because you are the only one who is frank with me."

"It takes courage to be candid with a prince. But, thank Heaven, I have it."

"Oh, pshaw! There's nobody by to admire your sarcasms."

"I'm satisfied with you, my dear boy."

"Will you answer me a serious question seriously?"

"Yes, if you keep your hands off, and don't try to pat me on the head."

Langbrith was silent, and he would not speak, in his resentment, till Falk said, "Fire away."

Still it was an interval before Langbrith recovered poise enough to ask, "What do you think of Jessamy Colebridge?"

"Hope Hawberk, you mean," Falk promptly translated.

Langbrith laughed, and said, "Well, make it Hope Hawberk."

"She's about the prettiest girl I've seen."

"Isn't she! And the gracefulest. There's more charm in grace than in beauty, every time."

"There is, this time, it seems."

Langbrith laughed again for pleasure. "She has grace of mind. I don't know where she gets it. Her father—well, that's a tragedy."

"Better tell it."

"It would take a long time to do it justice. He was my father's partner, here, when the mills were started, and I've heard he was a very brilliant fellow. They were great friends. But he must have had some sort of dry rot, always, and he took to opium."

"Kill him?"

"No, it doesn't kill on those terms, I believe. He's away just now on one of his periodical retreats in a sanatorium, where they profess to cure opium-eating. There's a lot of it among the country people about here—the women, especially. When Hawberk comes out, he is fitter than ever for opium."

"Well, that's something."

"I suppose it's Dr. Anther that keeps him along. I want you to meet Dr. Anther, Falk."

"I inferred as much from a remark you made at dinner."

"Oh, I believe I did speak of it. Well, now you know I mean it. He's one of those men—doctors

29

or lawyers, mostly; you don't catch the reverend clergy hiding their light under a bushel quite so much—who could have been something great in the larger world, if they hadn't preferred a small world. I suppose it is a streak of indolence in them. Anther's practice has kept him poor in Saxmills, but it would have made him rich in Boston. You mustn't imagine that he's been rusting scientifically here. He is thoroughly up to date as a physician; goes away now and then and rubs up in New York. He's been our family physician ever since I can remember, and before. My father and he were great cronies, I believe, though he's never boasted of it. I have inferred it from things my mother has dropped; or perhaps," Langbrith laughed, "I've only imagined it. At any rate, he dates back to my father's time, and two strong men, both willing to stay in Saxmills, must have had a good deal in common. He's always been in and out of the house, more like a friend than a physician. A guardian couldn't have looked after me better, when it was a question of advice; and, as a doctor, he pulled me through all the ills that flesh of kids is heir to. He has that abrupt quaintness that an old doctor gets. He would go into a play or a book just as he is. You don't care so much for that sort of man as I do, I know, for you're a sort of character yourself. Now, I'm different. I—"

"This seems getting to be more about you than your doctor," Falk said. He rose, threw the end of his cigar into the fire, and stretched himself.

"What is the matter with our going to see some of those girls?"

Langbrith flushed, as he rose too, but he said nothing in making for the door with his friend.

They met his mother before they reached the door, on her return from the kitchen. She gave the conscious start which every encounter with her son surprised from her since his home-coming, and gasped, "Will you—shall you—see the young people, James? Or shall I?"

"I can save you that trouble, mother. Falk and I were just going out to make some calls, and we can ask the girls."

"Well," his mother said, and she passed the young men on her way into the room, while they stood aside for her; she gave her housekeeping glance over it, to see what things would have to be put in place when they were gone. "Then, I will ask the others, and we will have the dance after supper. Were you going," she turned to her son with, for the first time, something like interest, "to ask Hope?"

"Why, certainly!"

"Yes. That was what I understood."

"Didn't you want me to?—I mentioned her."

"Yes, yes, oh yes. I forgot. And your uncle John?"

"Yes, certainly. But you know he won't come. Wild horses couldn't get him here."

"You ought to ask him."

"Now, that's just like my mother," Langbrith said, as he went out with Falk into the night. "Uncle

31

John has had charge of the mills here ever since my father died, and he was nominally my guardian. But he hasn't been inside of the house, I believe, half a dozen times, except on business, and he barely knows me by sight."

"The one I met yesterday in the office?"

"Yes. That's where he lives; that's his home; though, of course, he has a place where he sleeps and eats, and has an old colored man to keep house for him. He's a perfect hermit, and he'll only hate a little less to be asked to come than he would to come. But mother wouldn't omit asking him on any account. It makes me laugh."

VI

THE young men walked away under the windy April sky, with the boughs of the elms that overhung the village street creaking in the starless dark. The smell of spring was in the air, which beat damply and refreshingly in their faces, hot from the indoors warmth.

Langbrith was the first to speak again; but he did not speak till he had opened the gate of the walk leading up to the door of the house where he decided to begin their rounds. "Hello! they're at home, apparently," he said.

The windows of the house before them, as they showed to their advance through the leafless spray of the shrubbery, were bright with lamplight, and the sound of a piano, broken in upon with gay shouts and shrieks of girls' laughter, penetrated the doors and the casements. If there had been any doubt on the point made, it was dispersed at their ring. There came a nervous whoop from within, followed by whispering and tittering; and then the door was flung open by Jessamy Colebridge herself, obscured by the light which silhouetted her little head and jimp figure to the young men on the threshold.

"Why, Mr. Langbrith! And Mr. Falk! Well, if this isn't too much! We were just talking—

weren't we, girls?" she called over her shoulder into the room she had left, and Langbrith asked gravely:

"May we come in? If you are at home?"

"At home! I should think so! Papa and mamma are at evening meeting, and I let the two girls go; and I have got in Hope and Susie here to cheer me up, for I'm down sick, if you want to know, with the most fearful cold. I only hope it isn't grippe, but you can't tell."

She led them, chattering, into the parlor, where the other young ladies, stricken with sudden decorum, stood like statues of themselves in the attitude of joyous alarm which the ring at the door had surprised them into.

One of them, a slender girl, with masses of black hair, imperfectly put away from her face, which looked reddened beyond the tint natural to her type, flared at the young men with large black eyes, in a sort of defiant question. The other, short and dense of figure, was a decided blonde; her smooth hair was a pale gold, and her serenely smiling face, with its close-drawn eyelids—the lower almost touching the upper, and wrinkling the fine short nose—was what is called "funny." It was flushed, too, but was of a delicacy of complexion duly attested by its freckles.

There was a strong smell of burning in the room, and, somehow, an effect of things having been scurried out of sight.

The slim girl gave a wild cry, and precipitated herself towards the fireplace as if plunging into it; but it was only to snatch from the bed of coals a long-handled wire cage, from the meshes of which a thick, acrid smoke was pouring. "Much good it did to hustle the plates away and leave this burning up! Open the window, Jessamy!"

But Jessamy left Langbrith to do it, while she clapped her hands and stood shouting: "We were popping corn! The furnace fire was out, and I lit this to keep the damp out, and we thought we would pop some corn! There was such a splendid bed of coals, and I was playing, and Susie and Hope were popping the corn! We were in such a gale, and we all hustled the things away when you rang, for we didn't know who you were, and the girls thought it would be too absurd to be caught popping corn, and in the hurry we forgot all about the popper itself, and left it burning up full of *corn !*"

Her voice rose to a screech, and she bowed herself with laughter, while she beat her hands together.

The young men listened according to their nature. Falk said: "I thought it was the house burning down. I didn't know which of you ladies wanted to be saved first."

The girl who had run to throw the corn-popper out of the window came back with Langbrith, who shut the window behind her. "Oh, I can *swim,*" she said, and they all laughed at her joke.

"Well, then, get the corn, Hope," Jessamy shrieked; "we may as well be hung for a sheep as a goat. It *is* a goat, isn't it?" she appealed to the young men.

"It doesn't seem as if it were," Langbrith answered, with mock thoughtfulness.

"Some of those animals, then," the girl laughed

35

over her shoulder. "Where *did* I put the plates, Susie?"

"I know where I put the corn," Hope said, going to the portière, where it touched the floor next the room beyond.

Falk ran after her. "Let me help carry it," he entreated.

"Do get the salt, Susie," Jessamy commanded. "I know where the plates are *now*."

"We hadn't got to the salt," Susie Johns said; but Jessamy had not heard her when she stooped over the music-rack and handed up three plates to Langbrith.

Falk came with Hope, elaborately supporting one handle of the dish with a little heap of popped corn in the bottom. She held the other and explained, "We had only got to the first popping," and Jessamy added:

"We were not expecting company."

"We could go away," Langbrith suggested.

"Susie, *have* you got the salt?" Jessamy implored, putting the plates on the piano. Susie stood smiling serenely, and again the hostess forgot her. "Shall we have it on the piano, girls? Oh, I know; let's have it on the hearth-rug here."

"Yes," Langbrith said, doubling his lankness down before the fire. He went on:

"'For God's sake, let us sit upon the ground,
    And tell sad stories of the deaths of kings.''

Jessamy had not minded the hoyden prank in which he took her at her word, but the name he seemed to invoke so lightly shocked her. She drew her face down and looked grave.

"It isn't swearing, Jessamy," Hope Hawberk re-assured her; "it's only Shakespeare. Mr. Lang-brith never talks anything but Shakespeare, you know." She had a deep, throaty voice, which gave weight to her irony.

"Oh, all right," said Jessamy. "Susie, you wicked thing, have you got that salt? Why, of course! I never brought it from the dining-room. Here, sit by Mr. Langbrith, as Hope calls him—his Christian name used to be Jim—and keep him from Shakespearing, while I go for it."

"You might get him a plate, too," Falk called after her. Susie coiled herself softly, kitten-like, down on the rug at Langbrith's side. "*I'm* going to eat out of the dish."

"Hope, don't you let him!" Jessamy screamed on her way to the dining-room.

When she came back she distributed the plates among her guests, and with one, in which Hope had put her a portion of corn, she stood behind them. "Bless you, my children," she said. "Now, trot out your kings, Jimmy—Mr. Langbrith, I should say."

"Oh no," Langbrith protested; "ghosts. We oughtn't to tell anything less goose-fleshing than ghost-stories before this fire."

"Why, I thought you said your kings were dead. Good kings, dead kings!" Jessamy added, with no relation of ideas. "Or is it Indians?"

Anything served. They were young, and alone

—joyful mysteries to themselves and to one another. They talked and laughed. They hardly knew what they said, and not at all why they laughed.

At nine o'clock, Jessamy's father and mother came home, and with them some one whose voice they knew. The elders discreetly went up-stairs, when Jessamy called out to whoever it was had come with them, "Come in here, Harry Matthewson."

They received him with gay screams, Jessamy having dropped to her knees beside the others, for the greater effect upon the smiling young fellow who came in rubbing his hands.

"Well, well!" he said.

"Now this is a little *too* pat," Langbrith protested, and he gave the invitation which he had come with, and which met with no dissent.

"It is a vote," said Matthewson, with the authority of a young lawyer beginning to take part in town meetings.

"Well, now," Langbrith said, getting to his feet, "the business of the meeting being over, I move Falk and I adjourn."

"No, no, don't let him, Mr. Falk! You don't want to go, *do* you?"

"Only for a breath of air. I'm nearly roasted."

Matthewson laughed. "I wondered what you were sitting round the fire for; it's as mild as May out, and there's a full moon."

"A full *moon?*" Jessamy put out her hand for him to help her up. The other girls put out their hands for help, too. "Then I'll tell you what.

38

We'll go home with the poor things, and see that the goblins don't get them. What do you say, girls?"

"Oh! they say 'yes.' Don't you, girls?" Langbrith entreated, with clasped hand.

The young men helped them put on their wraps. Jessamy, when she was fully equipped for the adventure, called up-stairs to her mother: "Mamma, I am going out for a few minutes." Her mother shrieked back: "Jessamy Colebridge, don't you do it. You'll take your death."

"No, I won't, mamma. The air will do my cold good," and she closed the debate by shutting the door behind her. "Now, that's settled," she said. "Where shall we go first?"

The notion of going home with Langbrith and Falk seemed to be relinquished. They went about from one house to another, where there were girls of their acquaintance, and sang before their gates or under their windows. At the first sign of consciousness within, they fled with shrieks and shouts.

In the assortment of couples, Matthewson led the way with Susie Johns, Falk followed with Jessamy, Langbrith and Hope were paired. Sometimes, the girls ran on alone; sometimes, in the dark places, they took the young men's arms.

They saw each other to their houses; then, not to be outdone in civility, the girls who were left came away with those who had left them. It promised never to end, and no one seemed to care. The joy of their youth had gone to their heads in a divine madness, in which differences of temperament were merged and they were all alike.

Langbrith did not know how it happened that he was at last taking leave of Hope Hawberk alone at her gate. He stooped over to whisper something. She pulled her hand from his arm, and said, "Don't be silly!" and ran up the walk to her door. The elastic weight of her hand remained on his arm.

VII

THE compromise between a Boston dinner and a Saxmills tea, which the mother and son had agreed upon, prospered beyond the wont of compromises. It was a very good meal of the older-fashioned sort, and it was better served by Norah, from her habit of such meals, than could have been expected, with the help of the niece she had got in for the evening. The turkey was set before Langbrith and the chicken pie before his mother. Norah asked the guests which they would have, in taking their plates, and brought the plates back with the chosen portion, and the vegetables added by the host or hostess from the deep dishes on their right and left. There were small plates of subsidiary viands, such as brandied peaches and sweet pickles, which the guests passed to one another. Tea and coffee and cocoa were served through the supper by Norah's niece from the pantry, where she had them hot from the kitchen stove. There was no wine till the ladies left the table, when Langbrith had Norah put down, with the cigars, some decanters of madeira from, as he said, his father's stock. He had a little pomp in saying that; it seemed to him there was something ancestral in it.

Instead of letting all follow the hostess out to

supper pell-mell, as the Saxmills custom had always been, he went about asking the men, *sotto voce*, if they would take out such and such ladies. "Will you take out my mother, Dr. Anther?" he said, with special graciousness. He told Falk to give his arm to Hope Hawberk, and he gave his own to the rector's wife. But when they came to look up their places, and found their names, by Falk's example, on cards beside their plates, Hope found hers on Langbrith's left. That way of appointing people their chairs was an innovation at Saxmills, and the girls put their dinner cards where they should remember to take them away. But the effect of this innovation was lost in the great innovation of having old and young people together at tea. The like had not happened in Saxmills before; except at a church sociable or a Sunday-school picnic, it had scarcely happened that the different ages met at all. When they did, it was understood that the old people were to go away early, and leave the young people to take their pleasure in their own fashion.

At first, the affair went hitchily. The girls had confided to one another, in the library, their astonishment at finding themselves in the mixed company, and their wonder whether their elders were going to stay for the dance. But, partly through their fear of Langbrith, which they could overcome only when they had him on their own ground, and partly through their embarrassment at being obliged to talk with the rector and the doctor and the judge, they remained in a petrified decorum which

lasted well into the supper. Even when Jessamy
Colebridge caught the eye of Hope Hawberk from
her place diagonally across the table, and saw its
lid droop in a slow, deliberate wink, instead of bursting into a whoop of sympathetic intelligence, she
blushed painfully and turned her face away, with a
tendency to tears. She was not having, as she
would have said, a bit good time, between the judge
on one hand, who did not speak much to any one,
and Mr. Matthewson on the other, who was talking
to Susie Johns. And she felt the joyous mockery of
Hope's triumph, where she sat between Falk and
Langbrith, without the ability to respond in kind.
Besides, she could not see why her father and mother had not been invited, if there were going to be
old people. She could not catch the words which
were kindly cast her across the table, from time to
time, by the judge's wife. But good cheer is a
solvent which few spiritual discomforts can resist.
Before she left the table, Jessamy was beginning to
have the good time which mounted as the evening
went on, and culminated in Mr. Matthewson's going
home with her. Judge Garley had scarcely talked
to a young girl since his wife had ceased to be one.
But he was so little versed in the nature of girls
that he did not know how much he had failed to enjoy Jessamy's conversation till his wife asked him
at home how he could manage to find things to say
to that little simpleton. In fact, he had set her and
young Matthewson talking across him, while Susie
sat placidly silent, or funnily smiled to her indirect
vis-à-vis, who happened to be Falk, released to her

43

by Hope's preoccupation with Langbrith. As he noted to Susie, those two seemed to be having rather a stormy time, springing from a radical difference of opinion upon a point of sociology advanced by Langbrith, who held that the unions ought to be broken up, and alleged their criminal incivism even in their strikes in such a small place as Saxmills, where labor and capital were personally acquainted.

Mrs. Enderby was heard saying affably, across the table to Hope: "I didn't know young people took such an interest in those things. You ought to talk with Mr. Enderby. I'm afraid he finds me very lukewarm."

"Oh, well, then, *I'll* talk with you, Mrs. Enderby," Langbrith promised. "There's nothing I like so much as lukewarmness on these subjects. I'd no idea I should get into such hot water with Miss Hawberk. I believe she's a walking delegate in disguise!"

"Well," the girl said, "I shouldn't like anything better than to lead your hands out on a strike. I think it would be fun."

Mrs. Enderby said "Oh!" in compliance with the convention that one ought to be shocked by such audacity, but really amused with it.

"You'll find me in the ranks of labor, if you ever do lead a strike," Langbrith said, gallantly deserting his colors.

Hope went on: "I should like to be a great labor leader and start a revolution."

"What salary would you want?" Langbrith asked.

"About half the profits of the employers!" the girl came back.

"Well, we must talk to Uncle John about that. He manages the mills. But if your strike cut the profits down to nothing?"

"There, there!" Mrs. Enderby interposed. "You mustn't let your joke go too far."

"Oh, I haven't been joking," Hope said.

"I was never more in earnest," Langbrith followed, laughing.

His laughing provoked her. She wanted, somehow, to turn their banter into earnest—to say something saucy to him, something violent; something that would show Mrs. Enderby that she was not afraid of him. At the same time, she believed she did not care for Mrs. Enderby or what she might think, and in the midst of her insurrection it seemed to her that he was handsomer than she had ever supposed—that he had beautiful eyes. She noticed, for the first time, that they were gray, instead of black.

"How do you like my flowers?" he asked, as if their talk had been of the decorations of the table.

"Oh, did they all come out of your conservatory?" she returned, with an amiability which she could not account for. "It looks very pretty from here." She glanced down the table, with a quick turn of her little head, towards the glass extension of the room, where the plants bedded in the ground showed their green and bloom in masses under the paper lanterns, and the fine spray of an inaudible fountain glimmered.

45

"Yes, doesn't it? Everything that my mother touches flourishes."

"Oh, I know that!" the girl said, with an intonation of wonder and reverence.

"There are very few things," he said, from his proud satisfaction, "that my mother can't do better than anybody else."

"Did you have to go to Harvard to find that out? Everybody in Saxmills knew it!"

"But you haven't," he reverted, "said what you thought of the arrangement." He indicated the flowers on the table with a turn of his head.

Another mood seized her. "You can't spoil flowers!"

"Well, I did my worst." He wished her to know that he had suggested their arrangement.

Mrs. Enderby was talking with her left-hand neighbor. Langbrith lowered his voice slightly in asking: "Are you going to give me the first dance, Hope?"

"I don't know," she said vaguely; and then indifferently, "I suppose I must begin by dancing with somebody."

He laughed and they were silent, while she kept herself from panting by drawing each breath very slowly and smoothly. Her breast heaved and her nostrils dilated.

There came a quick clash on the glass roof of the conservatory. "Rain?" she said. "Goodness! How are we going to get home?"

"Oh, don't even talk of going home," he implored, and she laughed.

46

He looked down the table to catch his mother's eye, and give her the sign for rising with the ladies and leaving the room. That was a main part of his innovation and a thing unprecedented. But he had agreed with Falk that the stroke could be broken by each giving his arm, in the new fashion, to his partner, and taking her back to the library. The other men did not understand, and waited, on foot, for the cue from him. He lost his head, which seemed to whirl on his shoulders, and he was stooping to offer his arm to Hope when he remembered Mrs. Enderby.

He was stupefied into the awkwardness of saying, "Oh, I *beg* your pardon!"

The rector's wife laughed, from a woman's perennial joy in the sight of such feeling as his. "Oh, I shouldn't have minded."

Hope gave an imitation of not having noticed, which none but a connoisseur could have distinguished from the genuine.

VIII

"Dr. Anther, I want to introduce Mr. Falk a little more particularly to my oldest and best friend."

"Will he know what to do with such a treasure?"

Dr. Anther returned Falk's tentative bow with smiling irony, while he reached with his left hand for the cigars which Langbrith offered him.

Every one was still on foot, after leaving the ladies in the library, and Langbrith said to the group: "Sit down, gentlemen," and placed himself before answering the doctor. "Yes, I think he will. You smoke, don't you, Dr. Enderby? And you, Judge? Matthewson, I know, doesn't. Start the madeira after the sun, Harry." He pushed the cigars towards the elders and the decanter towards the young man, whom he bade give the smokers the candle. "Yes," he went on, to put Falk and Dr. Anther at ease with each other, "Falk's father is a physician, and my physician is the only father I have known."

"Oh, you're very good, James!" the doctor said, forgiving to the genuine feeling in the young fellow's voice the patronage of his words: "I can't say less than that no son of mine has given me less trouble." The two laughed together, and Falk smiled conditionally, as if he suspected that this

48

country practitioner had his knife out. "Are you going in for medicine, too?" the doctor asked.

"Worse yet," Langbrith answered for him. "He's going in for art. I don't know whether my mother has shown you any copies of *Caricature* which I send her. But, if she has, you know Falk's work. It's the best part of Falk. Falk is *Caricature* himself— with my poor help in joking a picture now and then."

"This puts me on my good behavior at once," Anther said. "Mr. Falk may be looking for types."

"No, no; Falk's types always look for him," said Langbrith. "Won't you sit down?"

"I've been sitting," Anther said, and he walked, Falk with him, towards the conservatory.

"Well, it's a change, and your smoke will help the plants," Langbrith called, and he turned to take part in the talk of the judge and the rector, to which Matthewson was listening with the two sorts of deference respectively due to the law and to the church.

"Well, Mr. Falk," Anther said, "I suppose we must make the best of being two such remarkable people. I hope you're enjoying your visit to Saxmills."

"Oh, very much," Falk answered, smiling less conditionally.

"I don't know that it's much adapted to pictorial satire."

"You must make allowance for the stately layout Langbrith gives his friends," said Falk, and the gleam of intelligence in the doctor's shaggily pent-roofed eyes satisfied him of his ground.

4   49

"The place has always struck me as very pictu-resque," the doctor continued. "Of course, I don't know; but a good head of water seems to imply broken ground, and if there's a fall, such as we have here, it means up and down hill and the broken banks and the rapids and other things that you ar-tists are supposed to care for."

"I don't know whether we really do—or I do," the young man said, modestly. "I'm rather more for the figure, I reckon."

"Western?" the doctor asked, with a lift of the pent-roofs.

"Northern Kentucky; Catletsburg."

"Curious! I thought of settling in that place, myself, before I came to Saxmills. Not New Eng-land people?"

"No, my people were German. My grandfather came out after the 1848 revolutions."

"Oh, indeed! Rather odd I should meet some one from Catletsburg at this late day. I've hardly thought of it since I gave up going there. Except for a run to New York, at times, I have been twenty-two years in Saxmills, and I don't suppose I shall ever go anywhere else to live. In that time, a man's life shapes itself to the environment, and new sur-roundings hurt. Don't you find it so?"

"Well, I'm just trying my first twenty-two years."

"To be sure," the doctor laughed. "I suppose you and James are thrown a good deal together at Harvard."

"This last year, yes. Since he took the editor-ship of *Caricature*."

"Oh, indeed! He must be very popular, then, to have that?"

"Not very," Falk answered, tranquilly. He looked steadily at the doctor, in breaking off his cigar ash, as if asking his eyes how far he might go. Then he said, in a low tone, but with a certain indifference as to being overheard in his manner: "A good many of the fellows think he's an ass. They can't stand him. But they make a mistake. He's got a lot of ability. He doesn't do himself justice."

"How?" the doctor asked, blowing his smoke out.

"Too patronizing. But he doesn't mean anything by it, as I know. All you have got to do is to call him down. He stands that first rate, if he likes you, or if he thinks you are right. And he stands by his friends."

"I'm glad to hear so much good of him. Naturally, I'm interested in him, knowing his—mother so long."

Falk asked—from a feeling that the doctor had meant to say "family" rather than "mother"— "You knew his father?"

"Oh yes, but he died when James was a very little child."

"He seems to have a sort of ancestor worship for him," said Falk, with a slight amusement in his face.

"Yes," the doctor dryly allowed.

Langbrith was talking gayly with his other guests at the farther end of the table, where his voice rose in somewhat noisy dominance. He seemed to be laying down the law on some point; and the others to be politely submitting rather than agreeing.

Anther stood looking at him. He turned to Falk, and, with his face slightly canted towards Langbrith, he asked from one side of his mouth, "You've noticed his portrait in the library?"

"Jimmy doesn't let you escape that!" Falk said.

"How do you like it?"

"You mean artistically?"

"No, personally. How does the face strike you?"

"Well, I don't think I could worship an ancestor like that. Perhaps it isn't a good likeness."

"It was painted from a photograph."

"Yes, so he said. And that sort of portrait seems always to fail in conveying character."

Doctor Anther made no reply for some time. In fact, he made no reply at all. He asked, "And such character as it does convey?"

"Well, he looks too much like a cat that has been at the cream. And it isn't the feline slyness alone that's there: there's the feline ferocity. Perhaps it's like a tiger that's been at the cream."

The doctor said gravely, "The artist had never seen Langbrith in life. You don't find anything of that sort in James?"

"No, *he's* like his mother in looks."

"Oh yes. Don't you find—as an artist—Miss Hawberk very striking?"

"Wonderful. If I may speak as an artist. That cloud of hair hanging over her little face, and those coal-black eyes, and that red mouth between the pale cheeks! If I were a painter, which I'm not, and never shall be, I should want nothing better than to spend my life studying such a face."

"Her father," the doctor said, looking at Falk, as if to question how much he knew already, "is an extraordinary man."

"So Langbrith tells me. He told me about him."

"Oh, then," said the doctor with the effect of implying that there need be nothing more said on that point; "you must stop me when I seem to be asking unwarrantable things. Do you think that James—"

"Doctor! Won't you come here?" Langbrith called to him from the other end of the table where he was sitting. "I've got three stubborn men against me here, on a point which I want you to settle in my favor."

"Somebody must give way, and you know I can't," the rector said, using the well-known words of the Boston lady who appealed to reason against her adversaries.

"What is this point that only one of you can agree on?" the doctor asked, coming up.

Langbrith laughed with high good humor, as if still in the afterglow of Hope Hawberk's playful hostilities. The qualities which gave his classmates the question whether he was not an ass were in abeyance. Even if he showed no such deference as Matthewson paid to the judge or the clergyman, he was withheld from patronizing them by the instinct of hospitality. At the worst, his superiority took the form of pressing the wine on them, and insisting that they had failed to get good cigars.

"Oh, I expect there will be two, now, doctor," he crowed. "It's a question of taste. I don't

53

know how we got to talking about it—do you, judge? or you, Dr. Enderby? But we were speaking of that immediate acquiescence of a community in a change of name—like that of Groton Junction to Ayer Junction. The pill-man gave a town-hall, or something like that, to the place, and the bargain was struck. Said, done: and from that day to this nobody has mentioned Groton Junction even by a slip of the tongue, though the school at Groton keeps the old name alive and honored. The judge, here, and Dr. Enderby were saying it was a pity that we had to keep such an ugly and indistinctive name as 'Saxmills,' and I was defending it, just because it was ugly and indistinctive. I was saying that the whole American thing was ugly and indistinctive, and that, if there was any choice, it was more so in New England than elsewhere. But now I want to tell you all something," and he went eagerly on, as if to forestall any interruptive expression of opinion from the others on a point which did not really matter. He glanced at Falk, where he stood blowing rings of smoke into the air at the door of the conservatory, as if about to demand his nearer presence, but apparently decided to include him by lifting his voice. "There was a time when a change of name was suggested here. Did you ever know about it?" he asked the doctor.

The doctor shook his head with indifference.

"No? Well, that was just like my father, if I read his character right. He would have consulted with you, if he had not decided of himself to suppress the whole thing from himself, and by himself. It was after he had built the library, and given it to the town. There was a dedication, and all that; and in a little diary—one of those little pocket-almanac diaries, you know—which I came across the other day among my father's papers, I found this laconic entry: 'Library dedication. Had been some talk of changing Saxmills to Langbrith, but I squelched that so thoroughly that nobody peeped about it.' Do you recall any such talk?"

Anther shook his head again. "It was before my time, here."

"And mine," the judge said.

The rector did not think it worth while saying it was before his, apparently; he was such a new-comer. But he said: "It was almost a pity he squelched it. Langbrith would have been a fine name."

The young man could scarcely conceal his satisfaction. "Oh, it would have been rather too romantic and Old English for a New England paper-mill town. I'm afraid it would have given the expectation of laid note, with deckelled edges."

The rector owned that there might be something in that, but he insisted that the name was fine.

"I think my father was right. And it was like him; don't you think so, doctor?"

"Very," Anther assented briefly.

"I can imagine just how he would have squelched it when the committee—there must have been a committee—came to propose the new name to him. I should not have liked to be in their shoes. He was not a man, as I imagine him, to have stood anything he considered nonsense." Langbrith look-

55

ed at the doctor for confirmation, but Anther smoked on in silence. Langbrith was probably too well pleased to note his silence with offence. He asked abruptly, "Is that a good likeness of him in the library?"

"It was painted from a photograph, you know."

"Yes, I know. Still, if it's well done, it would convey his personality."

"It's fairly—characteristic."

Falk, from his station between the conservatory doors, grinned.

Langbrith frowned slightly. "It doesn't suit me, quite. And this brings me to something I want to talk with you gentlemen about. I've been thinking, for some time, of offering the town a medallion likeness of my father to be put up in front of the library somewhere." He looked round at the others, but they waited as if for him to develop his idea fully. "My notion is, something in bronze; a low relief, of course, and a profile, or three-quarters face. The difficulty is about getting that view of him. The thing in the library is a full face, and I don't feel somehow that it does him justice. Do you, doctor?"

"Not perfect justice, no."

"He had a very strong character, but that painting conveys the notion of hardness rather than strength. Perhaps the hardness was something in the painter's method, and he couldn't eliminate it from the likeness." The judge and the rector smiled. Anther said nothing.

"But if I could get hold of the right man to do the

work, and could have you to help out from memory, doctor—"

"I couldn't," Anther said, abruptly.

The door-bell rang. Langbrith lost the frown in which his forehead had gathered, and smiled as he rose, and threw on the table the napkin he had been dragging across his lap while he talked. "There they come! This is something I should like to talk over with you gentlemen again." The judge and the rector made murmurs of friendly assent in their throats; the doctor did nothing to signify his acquiescence. "But, in the meantime, I would rather you wouldn't speak of it out of your own circle. Shall I follow you?" He made a motion for his guests to precede him, and called over his shoulder to his friend, "Come along, Falk."

IX

THE dance was coming to an end, and the girls, some of them, followed by as many young men, strayed out between the waltzes into the conservatory, to escape the heat; after trying the air, they said it was no cooler, only damper, and rushed back at the first strain of the music for the last figure of the dance. Hope Hawberk stayed, and Langbrith stayed with her. "Why don't you go back and look after your guests?" she challenged him.

"The guest that needs looking after most is here." He broke a rose from the vine at his hand, and threw it across the little fountain at her, where she stood with her head framed in the pale greenery of a jasmine bush. She lifted herself, haughtily. "May I ask what you mean, Mr. Langbrith?" Suddenly, while he stood, mystified and sobered, by the severity of her tone, she brought one hand from behind her, where she had been keeping both, and dashed a rose in his face. She tried to escape by the path that led up to the dining-room door past the callas in the oval bed about the fountain. He was instantly there to meet her, to catch her by a slim wrist and hold her fast.

"You witch!" he panted. "Oh, Hope, may I go home with you? The way we used to?"

"Before you were such a great person?"

"Why do you say that to me?" he entreated.

"Because—because you are hurting my wrist," she answered, with a child's wilful inconsequence.

He released it with all but his thumb and forefinger, and bent over it as if to see what harm he had done, while she stood passive. He kissed the red marks his fingers had left.

"What next, Mr. Langbrith?" she said, with a feint of cold impersonality.

"You know! Will you let me go home with you?"

"You're making me break your mother's lilies!"

"I don't care for the lilies. I care for you, you, you! May I go home with you?"

Another dash of the fitful April rain, which seemed to have gathered again, smote the glass roof; then it began to fall steadily. "You may lend me an umbrella," she said.

"Well, if I may go along to carry it."

"Oh, if you're afraid of not getting it back!"

"Yes, I can't trust you."

"You're hurting me again. Don't make me cry. Everybody will know it," she pleaded, releasing her wrist and passing her handkerchief over her eyes, with her face turned from the doors.

"Ah, Hope!" he tried to catch her hands, but she whipped them behind her, the handkerchief still in one of them, and ran, while he followed slowly.

The rain stopped again, before the dance was ended. The old people had gone home before, and the dancers now sallied out together into the air that had softened, since nightfall, under a sky where the moon sailed in seas of blue, among islands of white cloud. The girls started chattering, laughing, with meaningless cries, massing themselves at first, and then losing themselves from the group, one by one, and finding their way homeward with the young men who seemed to fall to their share, each as by divine accident.

Langbrith and Hope Hawberk were the foremost to put a space between themselves and the others, and he pressed closer and closer to his side the hand she let lie on his arm. "Will you say it now?" he was insisting.

"No more now than ever. What good would it do, I should like to know."

"How delicious! All the good in the world!"

"Well, I shall not. Why should you want me to be engaged to you."

"Oh, if you'll only say you love me, we'll let the engagement go!"

"Thank you! Well, we'll let it go without my saying anything so silly."

"But I may say that I love you."

"Yes, so long as you don't mean it."

"But I do mean it—I do, heart and soul. Hope, can't you be serious? May I write to you from Cambridge when I get back."

"How can I help that? I suppose the mail will have to bring your letters!"

"But will you answer them?"

"Perhaps they won't need answering."

"Oh yes, they will. I shall ask questions."

"Well, I never could answer questions. That's the one thing I can't do."

"Then you don't want me to write to you?"

"What an idea! I thought it was you that were doing the wanting."

"And I may?"

"Well, you may write *one* letter."

"Oh, how intoxicating you are, Hope!" He tried in his rapture to put his hand on hers, but it had slipped from his arm, and she was flying up the path before him. He followed after a moment of surprise; but, because she was fleet of foot, or because she had that little start of him, or because he felt the chase undignified, he did not overtake her till she had reached her gate. The little story-and-a-half house, overshadowed by two tall spruces, under the shoulder of the hill, was withdrawn only a few yards from the street, to which the gabled porch at the front-door brought it a few feet nearer.

She put her hand, panting, on the gate, and he had his on her shoulder, laughing, when, with an instinct of another presence, rather than a knowledge, she turned vividly towards him, and put her hand to her lip. He checked his laughter, and at her formal "Good-night" he said, reluctantly, "Well, good-night," and faltered outside the gate which she shut between them.

"Won't you come in, Jim?" a voice called huskily from the darkness of the little portico, and before he could formulate his "Oh no, thank you, Mr. Hawberk, it's rather late," the figure of a man ad-

61

vanced from its shadow. Around this figure Hope faded into the shadow it had left.

"It's only nine," Hawberk said. "Come in, and we'll have a bottle of champagne together. I'm just up from Boston, where I've been passing a week with some of your father's old friends: gay people. I was out at Cambridge, where I met some of the college grandees. They gave me great accounts of you. I was coming round in the morning to see your mother. She'll like to know direct from the university authorities that you are regarded as the most promising man there. I've been looking after an invention of mine, that I've succeeded in getting into good hands in Boston, and that will probably give me more money than I shall know what to do with. Have you ever thought of parting with the mills?"

"I don't believe I have, Mr. Hawberk," Langbrith responded.

"If you ever do," Hawberk said, "let me know. I've had an idea of taking them over, lately, and the income from this invention of mine will enable me to run them as they should be run. Your father and I were pretty close together in their management, at the outset, you know."

"Yes," Langbrith assented, while he retired a few steps from the gate, on which Hawberk was now lounging. In the moonlight, Hawberk's face had a greenish hue, and his eyes shone vitreously.

"There is something fine about these gloomy autumn nights," he suggested. "I sold him the mills, you recollect, and it would be sort of evening things up if you sold them back to me. Yes, your father and I were great friends. He liked to go off with me in my yacht. We made the trip to the Azores, together. I think I was the first to own a steam-yacht in Boston. I lived most of the time in Boston, then: looked after the city end of the business. Often had your father down. I was always giving dinners, and he used to enjoy them. You and Hope been at the play? Fine company, I'm told. Pity we don't get them oftener in Saxmills."

"Ah—I think I must say good night, Mr. Hawberk." Langbrith moved a little farther away, backing. "It's rather late—"

"Is it?" Hawberk took out his watch and held it up to the moonlight. "Why, so it is! Nearly morning. Well, good-night." He did not offer to leave the gate, but remained lounging across it, while Langbrith turned and moved down the foot-path towards the village.

X

In the morning, the dissatisfactions which are apt to qualify the satisfactions of the night before made themselves felt in Langbrith. He had wanted to talk the satisfactions over with Falk, whom he found in bed, on his return from seeing Hope Hawberk home, with the disaster of meeting her father; but Falk was either sleepy from the fatigues of the evening, or cynical from the excess of its pleasures, and would not talk. He met Langbrith's overtures to a confidence with a prayer for rest, with a counsel of forgetting, with an aspiration for help in his extremity against him from the powers which he did not often invoke. Langbrith was obliged to go to bed himself, without the light of Falk's mind on the things which kept him turning from side to side till well towards morning. Then he slept so briefly that he woke to hear Falk still asleep in the next room, and went down alone to his breakfast.

He found his mother in the library ready to join him, and he said, rather crossly, that they would not wait for Falk, who would anyway not want anything but coffee. At first, it seemed as if he would himself not want anything else, but after he had drunk a cup he helped himself to the steak which his mother refused, and then to the rice-cakes, which

Norah brought in relays, till he said, "I sha'n't want any more, Norah," and then she ceased to bring them, and shut the door into the kitchen definitely after her in going out.

If Mrs. Langbrith expected her son to begin by saying something of the pleasure she had tried to give him the night before, she was destined to disappointment, less, perhaps, from his ingratitude than from his preoccupation. "Mother," he asked, in pouring the syrup over the last relay of cakes that Norah had brought, "do you know whether there was ever anything unpleasant between Dr. Anther and my father?"

She caught her breath in a way that was habitual with her at any sort of abruptness, and had a moment of hesitation in which she might have been deciding what form of evasion she should employ. Then she asked, "Why, James, what made you think so?"

"Something—nothing—that happened, or didn't happen, last night, after you left us smoking in the dining-room." Langbrith frowned, in what was resentment or what was perplexity. "It might have been my fancy, altogether. But he seemed to receive a suggestion I made very dryly, very coldly. I had always supposed they were great friends."

Mrs. Langbrith quelled her respiration into long, smooth under-breaths, and said nothing.

Langbrith went on. "I had been thinking of something I meant to mention to you first—putting up a medallion of my father, with some sort of inscription, in the façade of the library, and last night I happened to come out with the notion in the course of some general talk, and Dr. Anther received it so blankly that I couldn't help feeling a little hurt."

"Perhaps," Mrs. Langbrith said, with a drop of her eyes, "he didn't take it in."

"That was what I have been trying to think. People began to come for the dance just after that, and the subject couldn't go any further. But, before Judge Garley and Mr. Enderby, Dr. Anther's blankness had time to be painful. Well!" he broke off from the affair. "He may not have taken it in, as you say."

Mrs. Langbrith rubbed her hand nervously up and down on the smooth, warm handle of the coffee-pot, in the struggle with herself, rather than with her son, which was renewed whenever it came to any sort of question of his father between them. She was long past the superstition of her husband's right, through the mere fact of his death, to her silence, her forbearance. Except for their son, she would have been willing that he should be known to the world as he was known to her and to Anther. But with reference to the dead man's son, it still seemed to her that the truth would be defamation, as much as if his memory were really pure and holy. It always came to some sort of evasion. But this morning, somehow, it did not seem to her as if she could consent to that any longer. It was on her tongue to say, No, his father and Dr. Anther were not friends at last, and give, swiftly and unsparingly, the reasons why they could not be. But when

66

she spoke, she got no further than saying, and it was with tremendous effect that she got so far from her wonted reserve, "If you think there was ever anything unpleasant between them, why don't you ask Dr. Anther himself?"

There was a desperate challenge in her eyes, which she would have been miserably glad to have him see there, if only some counter of his would then push her past the silence which she could never traverse of herself alone. But he was looking down into his cup, and he did not see what was in her eyes. He stirred his coffee, and said: "It was not serious enough for that. Very likely it wasn't anything at all. He may not have been giving the matter close attention, or he may have had something else on his mind. Doctors often have, I suppose; or he may have been vexed at something in my manner—what Falk calls my patronizing. Possibly he was thinking from his knowledge of my father that such a thing would be distasteful to him. But he might have left it all to me. Well, it doesn't really amount to anything."

She drew a long, deep breath, in the desperate relief of postponement, and he looked up affectionately. "It's all a very old story for you, mother, and you can't take much pleasure in knowing how the evening went off. You did manage it wonderfully."

She flushed at his praise. "I tried to carry out your instructions."

"You bettered them. It was a great little triumph. Don't you think people enjoyed it?"

"Yes, I think so. But if you enjoyed it, that is quite enough for me."

"Oh, for you, mother! But I'm unselfish enough for you to wish the rest had a good time. I thought the girls all looked very pretty, and they behaved prettily, too, which doesn't always follow. Country girls—village girls—don't always know the difference between being lively and being rowdy. I'm bound to say that sometimes city girls don't either. The latest blossoming of buds in Boston—well! Don't you think Hope is very beautiful?"

He seemed quite in good humor, now, and was smiling retrospectively. His mother said, from that remote caution, doubtless, which is in every woman where her son's relations with other women are concerned, "She is a very good girl."

Langbrith laughed out. "Well, I wasn't thinking about the goodness, exactly! But I dare say she is good. What I'm sure of, though, is that she's stunning. Mother!"

"Well, James?"

Langbrith's face, so like her own face, in its contour and features, flushed as hers always did with any strong feeling; but whatever his feeling was, he did not put it into the words which followed as from a second impulse. He gave himself time to lose his flush, and to knit his brows, which approached very nearly together, before he asked, "How long has her father been an opium fiend? I mean, how long have people known that he eats opium?"

"A good many years, I'm afraid."

"As long back as to my father's time?"

68

"Yes—quite. Why, what makes you ask?"

"Oh, I saw him last night when I went home with Hope."

"I thought he was away at the Retreat."

"It seems not. At any rate, he was at home, and she didn't seem surprised at his being there. It isn't like alcoholism, is it? It doesn't make him violent? So that he ever hurts them?"

"Oh no, not at all. Did Hope seem troubled?"

"No. She slipped into the house behind him, when he came out to the gate to talk to me. He was disposed to be rather expansive. Just in what way do you understand that he has been an affliction to them?"

"He has kept them poor."

"Well, that might be remedied. And it isn't the worst thing that could happen. A great many people are poor and happy. You don't mean that they're ever in anything like want?"

"Oh no," Mrs. Langbrith sighed. "He has some of his inventions in the hands of other people, who pay him a percentage on them, and it is secured so that it goes to his family, instead of to him. The worst of him is that they can't put the least dependence on him. They can't trust anything he says. He is very kind to them when he is with them, and he is proud of Hope. But they can't believe a word from him."

"He got off twenty inventions to me, in as many sentences, while we stood talking over the gate. I had a notion of something of the kind you say. Doesn't he ever blunder into the truth? He said

69

my father and he used to be great chums. Was there nothing in that?"

"They were friends at one time, certainly."

"Until he began to give way to all kinds of invention. Then, of course, it had to come to an end. Well, it's interesting to know that he can sometimes make a straight statement. Don't think I don't feel the awfulness of it, mother. I do, and I pity Hope, and I can understand how she can't help thinking that she is put wrong by it with—people. I suppose it's that that makes her a little defiant, a little doubtful of— Have you ever, or has she ever, mentioned the subject?"

"Not to me, James, or to any one that I know of. Everybody knows it. It's an old thing, and nobody talks of it, except new-comers. And there are not many new-comers here."

"No," Langbrith assented, with a smile. "Saxmills is static."

His mother may not have known just what he meant, or it may have been from the country habit of making no comment in response to what was not a question. She asked, "Will you have some more coffee, James?"

"No; but have them keep it hot for old Falk."

"I will have some fresh for him."

"There never was such thoughtful hospitality as yours, mother," Langbrith said, rising and going round the table to where she had risen too, and putting his arm fondly across her shoulders. She was almost as tall as he, and their likeness showed as he laid his face against hers and rubbed his cheek on

70

her own. "I believe that when I wake up in the other world you will be there to offer me something nice to eat. Old Falk is having a tremendously good time, don't you think?"

Mrs. Langbrith said, "Everything has been done for him that could be, by everybody."

"And I'm glad it's happened to Falk, too. A great many of the fellows don't know what a good fellow he is. They don't get hold of him. Falk is proud, and that makes him shy. Last year I wouldn't have thought of bringing him here, or getting him to come here. His people out in Kentucky are Germans, and they've always gone with the Germans. If Falk hadn't come to Harvard, he never would have got into American society. Fellows from out that way, where the Germans are rather thick, say that the third generation gets in, and sometimes the second if the first has got rich. But Falk's father is only a very musical doctor with a German practice, and no social instincts or aspirations. Of course, it's Falk's work in *Caricature* that's brought him forward with the best fellows. He's going to be a great artist, I believe, and I want to have a hand in helping him. It's difficult. He would rather say a nasty thing than a nice thing to you, and that doesn't cement friendship with everybody. But the way is not to mind it. He's all right at heart, if he wasn't so proud."

"I don't think it's very polite," Mrs. Langbrith ventured.

"Well, no," her son owned, "but it's better than being slimy."

7I

XI

LANGBRITH and his friend took the Northern Express in the afternoon, which would bring them to Boston just in time for dinner. Mrs. Langbrith gave them such a heavy lunch that, what with the sleep they had still to make up from the night before, they drowsed half the way to town in the smoking-car, which they had to themselves until the train began to stop at the suburban stations. Before this happened they woke, and Falk took a sheet of crumpled paper from his pocket, and spread it on the little stationary table between them which the commuters used for playing cards.

"How would that do for the next cartoon?" he asked.

He pushed it towards Langbrith, who smoothed it out again, and examined it carefully. "I don't know what it means," he said, at last.

"Neither do I," Falk said. "I want you to joke it, so that I shall."

Langbrith continued to look at the drawing, but apparently with less and less consciousness of it. He returned to it in pushing it away. "I don't know that I feel much like joking, to-day."

Falk crumpled the drawing up in his hand and threw it on the floor. "There oughtn't to be any

to-morrows. There ought to be nothing but yes-
terdays. Then we could manage."

"What do you mean?" Langbrith demanded.

"You're thinking you went too far."

"How do you know that?"

"I saw you going."

They were silent, and then Langbrith said, with a
laugh, "Well, if I went too far, I wasn't met half-
way."

"He laughs bitterly," Falk interpreted. "He
has got his come-uppings."

Langbrith looked angrily at him. Then his look
softened, if that is the word, into something more
like sulking than anger, and he said, "Sometimes I
think you hate me, Falk."

"No, you don't. You merely think you deserve
it. What have you been doing? You might as well
out with it now as later; I don't want you coming
in to-night when I've got into my first sleep."

"If I could only hope to make you understand!"
Langbrith sighed. "It isn't merely our having known
each other since she and I were kids, and always
been more or less together. And it isn't the country
freedom between fellows and girls. You could appreciate both those things. But you're so con-
foundedly hard that you wouldn't see why I should
feel a peculiar tenderness—a kind of longing to
shield her and save her: I don't know!—when I
think of her home life, and what it must be. I
know what a brave fight she puts up against its
seeming any way anomalous, and that makes her
all the more pathetic. It makes her all the more

73

fascinating—to a man of my temperament. She knows that, and that is why she is so defiant. I never knew she was so beautiful till this time. Weren't you struck with it yourself, Falk?"

Falk nodded, and smoked on.

"The complication of qualities in her, and the complication of her conditions, are what make it impossible to decide whether one has gone too far or not. Her way of taking it doesn't help you out a bit. She takes everything as if you didn't mean it. Of course, she knows that I'm in love with her. Everything I do tells her so, and so long as it isn't put into words, it seems all right. But when it comes to words, she won't stand it."

"She threw you down? Is that it?"

Langbrith frowned, and then smiled, as if forgiving the slang that might well have offended against the dignity of the fact. He even adopted it. "Not just threw me down, I should say."

"What happened, then?"

"Nothing. But I was in the mood for making her answer something more than she would answer, and I shouldn't have left her without, if it hadn't been for her father coming on the scene. He was an element that I hadn't counted on, and he made the whole thing luridly impossible. He seemed to cast the malign shadow of his own perdition over her."

"Good phrase," Falk murmured.

"Oh, don't mock me, old fellow!" Langbrith implored. "Of course, his being what he is wouldn't make me give her up, though I believe it would

74

make her give *me* up. Poor wretch! You can't think how amusing he was, with the wild romances he got off to me by the dozen in the two or three minutes we talked together. Do you remember that wonderful liar in one of Thackeray's stories, or sketches, who says he has just come from the Russian embassy in London, where he had seen a Russian princess knouted by secret orders of the Czar? It was something like that. That fellow must have been an opium-eater, too. One good thing about it," Langbrith resumed, after a pause not broken by Falk, "my mother thinks the world of Hope. She's always having her at the house, when she will come. I think she does it because my father was his friend in his better days, and she feels that he would like to have her do it. She is just so loyal to his memory. If she could imagine any wish for him, now, after twenty years, I believe she would want to carry it out, the same as if he were alive."

Falk still said nothing, and Langbrith broke off to say, "There was something that gravelled me last night, a little. I don't know whether you noticed it."

"What was it?"

"Well, Dr. Anther's snubbing way of meeting what I said of that medallion of my father which I suggested for the public library. It embarrassed me before the judge and Dr. Enderby; it made me feel like a fool. He had no business to do it. But, perhaps, he was merely not noticing. All the same, I'm going to do it. I think it's a shame that in a place which a man has done so much for as my father did for Saxmills there shouldn't be any public record of him. I'll do it to show them they ought to have done it themselves, if for nothing else. But I know all this bores you," Langbrith ended, vexed with his evident failure to interest his friend.

Falk yawned, but he said, with more than the usual scanty kindness he showed for the wounds of Langbrith's vanity, "No, no, I'm just stupid from last night. One doesn't have such a good time for nothing."

"It *was* a good time, wasn't it?" Langbrith gratefully exulted.

Falk said, "Fine." He yawned again, and Langbrith lapsed into a smiling muse, in which he was climbing the hill with Hope Hawberk, flattered in the fondness she suffered him to show her, and sweetly contraried by her refusal to say the words which would have sealed the bond between them. Was it, he wondered, with a swelling throat, because she wished to let him feel himself wholly free, in the event of some disgrace or disaster to herself from her father? He would live to prove that he would not be free: that he was hers as she was his, and nothing on earth could part them. That would make right, it would consecrate, all his past love-making. Once he would have thought that no harm, if it had come to nothing. But now, in his knowledge of another world, with a different code, it was not to be thought of but as part of a common future for them which it began. He wanted to put the case concretely before Falk, but he could not. He could not generalize, as he would have liked to do, on that difference of code between city and country, with the risk of Falk's making his abstractions concrete in some such way as only a blow could answer. Falk had his limitations. After all, he was only half an American, and he could only half understand an American's feelings. He retreated from the temptation, and lost himself in a warm revery of the future, which he forecast in defiance of every obstacle.

He thought what friends Hope and his mother had always been, and he knew that there could be nothing but glad response in his mother's heart to the feeling that was in his for Hope. Then he began to think of his mother apart from Hope, and of what she might have been like when she was a girl. She was younger even than Hope when she was married. She had been many more years a widow than a maid; and, in the light of his own love for Hope, he wondered if his mother had ever thought of marrying again. His father had been twice her age when he married her. Langbrith knew this in the casual way in which children know something of their parents' history, and his father must have been an uncommon man to have won her with that difference of years between them, and to have kept her constant to his memory so many years after his death. After all, how little she had ever said of him! Langbrith romanced her as not being able, from deep feeling, from a grief ever new, to speak of him, and he ached at heart to think how his father's personality seemed buried in his grave with his body. A tender, chivalrous longing to champion his forgotten father, to rehabilitate this vanished personality, replaced his heartache, and again he was indignant with Dr. Anther for his indifference, his coldness. He said to himself that he must have an explanation from Dr. Anther; he would write to him, and ask just what he meant. Perhaps he meant nothing. But he must be sure. Then he would see that young sculptor, that Italian, and tell him what he wanted; talk it over with him; find if he had any notions of his own.

The train slowed into the North station about five o'clock, just when he knew his mother would be talking with old Norah about the supper, to which, in his absence, she would revert from the late dinner. She would be bidding Norah tell the cook that she did not want anything but a cup of tea and a little milk-toast. Poor old mother! What a savorless, limp life she lived there alone! Yet it could not be otherwise, when he was away. How much she depended upon him! Somehow, he must manage for her to live with Hope and him. She must go out to Paris with them, where they should go after their marriage, and when they came back to Saxmills, where they would always have their summer home, she must be put back mistress in the old house.

XII

THE neighbor over the way who saw Anther drop
the hitching-weight of his buggy in front of the
Langbrith house, late in the afternoon of the length-
ening April day, decided that Mrs. Langbrith had
been overdoing. She watched for him to come out
until she could stay no longer at the window without making her own tea late, but she did not see
him come out at all.

In fact, it was the doctor who appeared to have
been overdoing. He looked so tired to Mrs. Langbrith that she asked him if he would not have a
cup of tea. Upon second thought, she asked him
if he would not have it with her. Supper would be
ready very soon; and, without waiting for a refusal,
she went into the kitchen to hurry it, and to have
the cook add something to the milk-toast for the
man-appetite, to which her hospitality was minis-
tering with more impulsiveness and spontaneity
than the wont of village hospitality is.

When they sat down together at the table, he
did not eat much and he talked little; but he seemed
to feel gratefully the comfort of the place and presence. She came into authority with him, as a woman does when the man dear to her is depressed. Her
affection for him came out in little suggestions and

79

insistencies about the food. Like most physicians, he kept his precepts for himself and his practices for his patients. He now ate rather recklessly, and he preferred the unwholesome things. At first she had to press him, and then she had to check him. At last she had to say to Norah, who came in with successive plates of the hot cakes which he devoured, "That will do, Norah," and, when he had swept the final batch upon his plate and soaked them in butter and syrup, and then cut their layers into deep vertical sections, and gorged these with a kind of absent gluttony, while she looked on in patient amaze, she rose and led the way from the table into the parlor.

It lay beyond the library and had windows to the north and east. The library was lighted from the east alone, like the dining-room in the wing. The main house was square, and divided by an ample hall from front to back. Beyond the hall, the two drawing-rooms opening from it balanced the parlor and library. There was a fire of logs burning on the parlor hearth, and its glow alone lighted the place when the two came into it. He went first to the window and looked at his horse. When he came away she pulled down the curtains and shut out what was left of the pale day and the disappointment of the neighbor who had been waiting for the reappearance of the persons of a drama not played for her.

Mrs. Langbrith took the chair at the corner, and invited Anther to the deeper one in front of the fire by her action.

"I oughtn't to stay," he said, looking at his watch. But he sat down. Neither of them made haste to take up any talk for the entertainment of the other. What they were to say was to come because they were both thinking the same things, from interests that were no longer separable. Yet he began with as great apparent remoteness as possible from their common interests. "Hawberk is at home again," he said, as if that followed from his saying he ought not to stay.

"James told me," she responded. "He saw him last night."

"And he has begun again."

"Yes, I knew that from the way that James said he talked. It doesn't seem much use his ever going."

"It prolongs his life, if that's any use. If he hadn't pulled up completely, from time to time, he would have been dead ten years ago. It is a curious case. Mostly they keep on and on, till they kill themselves, but Hawberk seems disposed to see how much relief can be got out of it with the least danger. At the rate he is going, he can live as long as anybody. Of course, the moral effect always follows the indulgence of a morbid appetite. What did he say to James?"

"He just told him some of his wild stories. He boasted of being Mr. Langbrith's greatest friend."

"So he was, in a kind of way. An involuntary friend," Anther said, with a smile. She smiled, too, strangely enough, but as people can smile, in dealing with an old wrong when it offers an ironical aspect to them. But she said, "Sometimes I wish it could be known what a deadly enemy Mr. Langbrith had been to him. Why shouldn't I tell it? I ought to feel guilty for not telling it. He robbed him, as much as if he had taken his money out of his pocket."

"No doubt about that; and once it might have been best to own the fact publicly. But sometimes it seems to me that time is past. A wrong like that seems to gather a force that enslaves those who have done nothing worse than leave it unacknowledged through a good motive. You haven't been silent for your own sake."

"I am not sure it hasn't been for my own sake."

"I am."

"I wonder," she said, "that Mr. Hawberk hasn't told it himself."

"Well, possibly, he thinks that it wouldn't be credited, that it would be regarded as one of his wild inventions; that is, he thinks that when he is in his soberer moments. When he is under the influence of the drug, he likes to make pleasing romances, and has no desire to mix a tragical ingredient in them."

"Then Mr. Langbrith has ruined a soul!"

"Yes," Anther admitted, "he has done something like that. And the most terrible thing is, that he holds the man in bondage now much more securely than he could have held him living. If they were both still alive, there would be some means of righting the wrong that has been done.

Some pressure could be brought upon him to make him do Hawberk justice."

"No, no, he would know how to get out of that." She rose and closed the door opening into the library. She had meant to do it quietly, and without self-betrayal; but, in the nervous stress that was on her, she brought it to with a clash, and then she felt obliged to explain: "It always seems as if it were listening," and Anther knew that she meant the portrait over the library mantel.

"At any rate," the doctor resumed, "he makes it hard for you to do him justice now. You do the best you can, and perhaps it is the best that any one could do. I suppose that a moralist, like Enderby, for instance, would say that the secrecy which Hawberk's misfortune promotes is the worst part of it. You pay Hawberk an income from a stolen invention, and he goes about bragging of the inventions which he has in the hands of Boston capitalists. Perhaps it is not even possible for him to tell the truth, in the perversion of his nature through his habit."

"What was he like before he took to it, Dr. Anther?" she asked, from the security she felt in shutting out the portrait. "I know that he took it up in the misery he felt at being trapped and robbed, and it was his only escape."

"Do you mean, whether he was inclined some such way?"

"I have sometimes wished that he were."

"He may have been," the doctor mused. "I knew him very little before I came here. But there is a sort of crime, isn't there, in pushing a man in the direction of a natural propensity? You don't want to palliate what was done?"

"Mr. Langbrith was capable of any crime," she answered. "Sometimes I have to shield his memory. But I don't wish to do it when I needn't. That is the comfort, the rest, of talking with you. I can't tell you what a kind of awful happiness it is to say out to you the things I cannot say to any one else. You will think I am crazy, but the next greatest happiness I have is in hoping that his fancy is taken with her, and that somehow it can be made up to them in that way. And yet there is a ghastliness in that, too, that is awful."

He knew that now she was talking of her son and of Hawberk's daughter. When she added, "She ought to know, at least," he said:

"Oh, everybody ought to know. But it is no more possible for her to be told than for any one else. I should be glad if he could get so good a girl. She is a beautiful creature, too, as well as good. Well!"

He rose from his chair, but from hers she entreated almost unawares, "Oh, don't go! Or, I oughtn't to say it!"

"No, Amelia, you oughtn't. If you said something else, I need never go." He looked at her sadly, and her head drooped. "You let me see an image of home, like this, and then you take it from me. Well! I must submit. Good-night." He put out his hand to her, but she would not take it.

She lifted her eyes to his, "You haven't asked me if I tried to speak to James. I didn't!"

"I knew that."

"Perhaps I should—perhaps I should have tried, this morning, when we were alone, if— But perhaps I couldn't."

"If what?"

"If he hadn't fancied that you did something last night that showed dislike of Mr. Langbrith."

"What was it I did?"

"Something in the way you received his suggestion of the memorial tablet."

"Oh, he noticed that? Well, I couldn't help it."

"I know you couldn't. Do you think I blame you?"

"I believe we don't blame each other, Amelia."

"And you don't feel hard towards me for not trying?"

"I didn't expect you to try."

"But why shouldn't we go on like this—the way we have gone on for twenty years? Why shouldn't you be just my friend as long as you live? We are not young, and we couldn't expect what young people expect of marriage."

"I expect a great deal more," he said. "You are solitary, and so am I. I have never had a home, and you could give me one. I have never had companionship at the time when a man wants it most, and you could be my companion. I want some one to talk to and to be silent to, when I feel the need of either. You could be my daughter, my mother, my sister. Why do you make me say these things to you?"

"Well, then, why not come and let me be it here?

85

Why not come and make this your home? I know
James wouldn't object. I believe he would like to
have you live with us. He has always been used
to you—" Anther shook his head.

"Yes, yes," she persisted. "We could give you
all the room you wanted in the house here, and you
could have Mr. Langbrith's office for your office,
out there by the gate. I have thought how it could
be done—"

"It couldn't be done, Amelia. The talk it would
make in a place like Saxmills!"

"There wouldn't be any talk. You have been
here so long, and you are so respected. You have
always been our doctor, and you have been in and
out here day and night. You are like one of the
family. You could come now, when Mrs. Burwell
is going to give up her house, and you will have to
go somewhere else, anyhow. It hasn't made talk
your living there with her all these years, and why
should your living here do it? Sit down now, and
let me tell you—"

She had put her hand unconsciously on his arm
and was nervously pinching the sleeve. He took
her hand away and held it in his own. "I never
think of Mrs. Burwell, nor she of me; but we
two would always be thinking of each other. It
wouldn't do, my dear, and you know it."

She broke out piteously, "I am so afraid of James!"

"Yes, I understand that, and I should be afraid
of him, too, if I came here to live with you, unless I
came as your husband. In that case, I shouldn't
be afraid of him."

86

"Ah, you hate him! I can see it by the way you say that. What shall I do?"

"Nothing, Amelia, except be reasonable. I don't hate your son; how could I? Of course, your fear of him stands in our way, but I am not at all sure that *he* does. He might have done so, a few years ago, but there is less probability that he would now."

"How do you mean?"

"He is more rational. He is of a nature that matures late; he is like you, in that, Amelia. That friend of his, that young man, told me how slowly James has won upon the liking and understanding of his college mates. They did not like him at first, but now, in his last year, they are beginning to value him, to make allowances for what repelled them, to see how he has changed, and to have an affection for him." In his gloss of Falk's laconic terms, Anther did not feel that he was misinterpreting his statement of Langbrith's Harvard standing; his mother eagerly accepted the version, and imagined it insufficient. "I say this," the doctor went on, "merely to illustrate my meaning. He is now at the age when the mind acts with an insight unknown to it before, and besides—" Anther broke off, and then asked, after a moment: "What reason have you for thinking that he is seriously taken with Hope? How is it different with them from what it has always been?"

"I don't know. Perhaps it is his being away, and then coming back and finding her changed into a new person. Girls change so suddenly at her age. If he had stayed at home, they might have gone on being boy and girl together always. But as it is— Perhaps it is partly the way I have seen him look at her—with a kind of surprise. And this morning, he spoke of her with so much— Oh, if it only could be, what a load it would take off my heart!"

"It would take the main obstacle out of our path, too," Anther responded. "He would judge you somewhat more from himself."

Mrs. Langbrith colored faintly, with a kind of shame, which he saw and resented.

"You think it isn't the same thing!"

"No," she owned. "How could I? It is as right for us, though it is different, as it is for them. But—"

She stopped, and even after he had said, "Well?" she did not go on immediately.

Then she shook her head, and added," It wouldn't get over the great obstacle. There would still be —Mr. Langbrith."

"Then," said Anther, harshly, "we must remove that obstacle, that incubus, ourselves. That man's memory mustn't be allowed to be a lifelong nightmare to you. You suffered enough from him when he was alive. We must tell James about him."

"I couldn't."

"Then you must let me."

She slowly turned her head away. "It's too late," she sighed.

"No. Now is just the time. Before this, it would have been too soon. While he was a child, you could not have told him; I understand that; and you had to let him grow up in the superstition of such a father as he imagines. But now he is old enough and strong enough to have his fetish taken from him. You owe it to him to take it. Put me out of the question entirely. I will never speak to you again of what I wish—"

"Oh, do you think that would be any help?" she lamented.

"At any rate, it is the boy's right to know the truth now."

"I always," she tried to evade him, "hoped that some accident—"

"But it never did. And it never will. That isn't the way of accident. It doesn't manage beneficent surprises."

"It is too late. I can't let you tell him the truth, and I can't myself. It must be covered up, more and more! It must be hidden forever. But there is something—something you might do, and you could do it."

"For you?"

"For him—for me."

"Of course, I will do it."

"I don't know. You could help him—help me. What harm would there be in your humoring the child?"

"How, humoring him?"

"Why shouldn't you encourage him—why should you oppose him in putting up that tablet? Or not that! Why should you be so cold with him about it?"

Anther walked out into the hall, and got his hat

and coat from the rack there before he spoke. "Amelia!" he cried with a sternness that he let die out of his voice before he added, "Oh, poor woman! That scoundrel has had power to corrupt even you, even now."

He opened the outer door, and, while she stood on the threshold of the parlor, with entreating hands stretched towards him, he closed the door behind him without looking back at her.

XIII

Mrs. Burwell came to call Dr. Anther to break-fast as soon as she heard him in his office. He had been up late overnight, and, with the fretful patience which had not failed her in twenty years of obedience, she had obeyed his instructions not to call him in such a case at the established hour of seven. His breakfast was always ready at seven, and it would have been some consolation to give him his breakfast cold, if he ever noticed whether it was cold or hot, but he did not, and she failed of this comfort. Among the reasons which had decided her at last to give up house-keeping and go to live with her married daughter in Nashua, the irregularity of Dr. Anther at breakfast would have been found first by any one who cared to study them, but it was one which she urged last upon the inquirer's attention. She said that it had been clearly agreed upon at the beginning, and that she was not one to take back her word.

He sat before his desk opening his letters, with his revolving bookcase by his side, and, in the long case between the two windows behind him, the pendulent skeleton which he had bought with his practice, from his predecessor. When the case was closed, it looked like a grandfather's clock in shape, and when it was open it still suggested the intimate relation of time and death. There was a table in the room, and over this were scattered medical periodicals, and other publications more suited to the taste and intelligence of patients waiting for his return when he was out. There were some hard chairs which did not invite their fancy from the stern realities of life by luxurious appeals to the senses.

"Lorenzo Hawberk's b'en here," Mrs. Burwell complained to the back of the doctor's bowed head. "He said he would call in again. I don't know but what you'll find your coffee pretty cold," she lamented further.

"I'll be there in a minute," Anther said, still without lifting his head, and, when he had quite finished with his morning's mail, he followed her vanishing into the hall without, and thence into the dining-room.

If he had been the sort of man to realize the order of facts to which any article of food belonged by its condition, he would have found not only his coffee cold, but his biscuit and his steak cold, too. But he was only vaguely aware of something wrong, as a child is when it is in discomfort, and his sense extended itself still more vaguely to an impression of the room, and of Mrs. Burwell herself. They were both severely neat, and they were both of the same material and spiritual spareness. Beginning with the hard little knot which Mrs. Burwell's silver-sanded hair was tightly drawn up into, away from her face, a more than classic temperance of ornament was characteristic of both. In her and in the room, everything was designed and disposed with a view to not catching dust. The clock on the mantel, supported by two Japanese fans, and the four prints on the four walls, representing severally the Lincoln Family at Breakfast, the Battle of Gettysburg, the long-extinct husband of Mrs. Burwell, and the United States Senate listening to the speech of Mr. Webster of Massachusetts in reply to Mr. Hayne of South Carolina, united with the sideboard, on which there was nothing that could not be shut away in its drawers the moment the breakfast things were washed up, in preserving a condition which not only would not catch dust, but in which there was no dust to catch.

"Did he leave any word?" Dr. Anther answered, not troubling himself to name Hawberk in his question.

"No, he just said he would be back; there was nothing particular the matter. I suppose he's begun again."

Hawberk's habit was so notorious in Saxmills that Mrs. Burwell felt it no violation of that other convention between herself and her tenant, dating from the beginning, like the agreement in regard to breakfast, that she was not to offer any sort of comment upon his patients, their characters, their ailments, or affairs. All the same, he snubbed her by his tacit refusal to enter into the case of Hawberk with her.

"Have you heard from your daughter again, Mrs. Burwell?" he asked.

93

"No, I hain't," she said, with an effect of being resolved to have no concealments. "But, as far forth as that goes, I don't know as I expected to."

"Then you are still decided to go to her?"

"Well, yes, I suppose I am," she said, as little decidedly in words as a woman well could.

"Supposing won't do," Anther pursued. "I must know whether you really intend to go or not, for I must find some other quarters if you do, and I want time."

"Well, then, I *am*. I suppose I said 'suppose' because I didn't want to seem to be hurrying you up any."

"You'll hurry me up if you don't give me due notice."

Mrs. Burwell's hard mouth and hard eyes joined in the adamantine response which she made. "I'm goin' to leave this house the first day of July, no sooner and no later, as far as I can humanly fix it."

"Oh, well, then," Anther said, "that gives me plenty of time to look about. I thought you were going in June."

"Well," she admitted, reluctantly, and without that bravado of frankness which she had shown before, "I did some think of goin' in June, and I did think I might as well stay the summer out here. It's full more comfortable than what it is in Nashua, with the heat, and it's easier to begin in a new place where you've got to be shut up a good deal, anyway, by beginnin' in the fall of the year."

"Yes, that is so," the doctor granted, and Mrs.

Burwell chose to read a sympathy into his words which they did not express.

"I presume that I shall feel the change, and I presume you will, some."

"Yes, I shall hate the moving."

"That's what I mean. And I wonder you want to move. Why don't you take the house yourself? It 'll be to rent when I give it up. You could keep your old rooms here, and get somebody in to do for you—I don't know but what Orlando himself could. He's real handy about a house, and he knows your ways—till you could get somebody to take the rest of the house. You could meal out; you're so irregular, anyway. I declare I feel bad about breakin' you up here, and I don't like to have anybody comin' in that I don't know."

"Thank you, Mrs. Burwell," Anther said to the part of her speech that demanded thanks from him.

"I don't, one bit," she continued, with the other part. "And still I don't want to have it, as you may say, layin' empty."

"No, it would be a certain expense, and you would get no return from it."

"Yes, and a house wears out faster when it's empty. I'd be willing to let it to anybody that would take good care of it for two hundred and fifty a year."

"That would be reasonable."

"Why, it wouldn't hardly more than pay the repairs and taxes," Mrs. Burwell urged. "I shouldn't expect to make anything on it, though goodness knows I need to, with everything as dear as they say it is to Nashua. I expect to pay good board to my daughter, though I presume I shall do enough about the house to make up without payin' anything."

"Well, I'll see about it."

"So do." Mrs. Burwell did not rise, but stretched her long arm across the table for the doctor's plate; she cleaned it into her own, and began to put the table in order for his uncertain dinner before he left the room. He went out of the side door upon a back porch, where Mrs. Burwell considered it neater to do certain parts of her housework than indoors, and more convenient for the disposal of peapods, squash seeds, and all kinds of cores and peelings, as well as those bits of refuse from fowls and butcher's-meat which she could throw to the hens, netted into their yard beside the stable, without having contaminated her kitchen with them. She preferred to work there not only in summer, but as far into the winter as she could bear the cold, and, wrapped up as to her head and shoulders, she defied the elements till after Thanksgiving and well towards Christmas. Her back yard, between this porch and the stable, was as clean as the front yard, which dropped from the terrace where the house stood, and sloped three yards and no more to the white paling fence in the gloom of four funereal firs, cropped upward, as their boughs died of their own denseness, till their trunks showed as high as the chamber windows. The house was painted of a whiteness which age had never been suffered to soften, but was as coldly fresh as the green of the shutters; it had been there thirty years, but it stood as prim and new to the eye in every detail as if it had been finished the week before. Mrs. Burwell herself never appeared in the front yard except to pick up the fir twigs dropped in the spindling grass that bearded the terrace; in the immediate shadow of the trees no grass grew, and the ground was matted with the dark-brown decay of their spray and spills, and looked as if it were burned over.

Dr. Anther noted that his buggy must have been driven to the front gate, since there were no signs of it at the stable door, and he walked round the house, and looked up at its frigid façade with a novel interest. It had been long since he had looked at it, though he had daily gone in and out, and had slept in the northeast chamber ever since he had been Mrs. Burwell's lodger. A certain shallowness of the structure now appeared to him, and he realized that the front was but one room deep on each side of the door, and that it shrank behind into the ell which imperfectly supported its pretensions of squareness, by stretching into an indefinite extent of kitchen and woodshed beyond the dining-room. He perceived that he had the two best rooms, but that the parlor, and the chamber above it, which was kept as a guest-room, though he could not remember when there had been a guest in it, were as large, if not as pleasant, as his own. From the fact of back stairs, he had always inferred a chamber over the dining-room, and he had conjectured something of the sort in the sloping roof of the kitchen. There were, then, eight rooms in all, and it did not seem to Anther, though he gave the matter no very distinct thought, that there were too many for the money that Mrs. Burwell proposed getting for it in a place like Saxmills. He dropped his cursory glance from the façade to the front door, and noted, with the sort of novel interest that the whole had inspired, the name of Justin Anther, M.D., on its small, glass-framed plate, and then he went in-doors.

There was some one waiting for him in his office, and he said, "Ah, Hawberk!" in greeting of the presence which he had inferred from the legs he had first seen, as they stretched across the perspective of the door-way.

The man got to his feet with a certain alertness, which was more like a reminiscence of past activity than an actual fact, and offered the doctor a wasted hand. He looked shrunken within his clothes, and his greenish-brown complexion, blotched with patches of deeper brown, where the skin showed above the lustreless beard, was lighted with eyes which were still beautiful, though their black was dimmed by the suffering through which they had sunk into their cavernous sockets.

"Good-morning, doctor," Hawberk said, and he added, courteously, "I hope I see you well?"

"I'm fairly well," Anther said, facing round in his swivel-chair, which he had taken at once. "Sit down, won't you? When did you get back?"

"Oh, I've been back some time—about a fortnight, I should say. But I've been pretty busy with a little thing of mine that I'm working at, and I haven't been about the town much."

"I hope my old friend Fredericks was able to do for you what you wanted?"

"Oh yes, oh yes," Hawberk answered, nervously, but with a vagueness that did not seem to belong with the quickness. "He set me up. I'm all right now."

"Gone back to it since your return?"

"Well, no. I can't say that exactly. Still I don't think it's well to make an entire break. I think the tonic effect is good, don't you?"

"Perhaps. If you don't make it too tonic. How much have you got back to?"

"Well, it ain't worth mentioning. Two or three spoonfuls after meals, and as many more at night."

"Dreams all they ought to be?"

"Oh yes, they're all right, now. I'm out of that pit that used to give me so much trouble. I don't have to keep digging at it the whole night now, as I used to before I went to the Retreat. Dr. Fredericks pulled me out of that fairly well. There is a small matter of old bones and a skull or two," Hawberk added, with a jocosity that did not make Anther smile. "But the great thing is that I understand it's a dream even while I'm dreaming it, and I guess I shall be able to break it up if I keep realizing it. And it doesn't seem to last so long. I think that's a decided gain, don't you?"

"It's not a loss," the doctor admitted.

"It's a fighting chance, and I'm taking all the fighting chances there are. I've fairly got the upperhand. If you were to tell me to leave the whole thing off, I could do it, and not turn a hair."

Anther made no answer, and Hawberk sank from his bragging note into a dull, confused tone, as he rubbed his hand over his forehead tremulously. "There was something I wanted to tell you about—"

The doctor prompted him after a moment's wait, "Anything about your condition?"

"No, no." As if he could not recall the thing he was groping for, Hawberk said, with a sort of provisionality, "I stopped in Boston on my way up yesterday, and saw that man who has my new patent in hand. He's a great fellow, and he's working it for all it's worth. He's sold territory over the whole country, and up into Canada. Why, doctor, he's got twenty thousand dollars for the Canadian rights alone, up to date, and I come in for a clean half of the money! I'm going to build, this spring. I've as good as bought that hill back of my house— got an option on it—and I'm going to build up there and keep the old place where we are for a shop. Have a walk slanting down to it over the corner of the hill, but have the main entrance to the new place by a flight of stone steps from the street. Have the whole front of the hill terraced. I've got a landscape architect in Boston studying it out for me. I was telling Jim Langbrith about it last night. He brought Hope home from the party at his mother's, and we got talking, and— Oh yes. Now I know what it is I wanted to speak with you about, doctor, It's a very confidential matter, and I don't know anybody that I'd like to trust with it except you. You at their house last night?"

"Yes," Anther owned.

"Well, all right. I couldn't go with Hope myself, for I had that man up from Boston that's handling my new patent. Had to send Mrs. Langbrith an excuse by Hope. But you saw them together, didn't you? And what did you think? Think there was anything serious? I mean in Hope. Because I know there is in him. He asked me last night if I had any objections to their getting married as soon as he is out of college. I haven't talked with Hope, and I don't know, except from him, how she feels."

Hawberk tried to fix Anther with the dull eyes that had once been brilliantly black and bold, but now seemed to slip in their glance, and he paused in the monologue which was like sleep-talking, a continuous babble, unbroken in its flow by the questions that interspersed it.

The doctor rubbed his chin and stared back at him. "Are you sure of what you say, Hawberk?"

"Sure!"

"Because, you know, you sometimes can't tell the facts from the dreams."

"Oh, but I can this time. I couldn't be mistaken about a thing like that. What do you advise me to do? I've got plenty of means to meet the Langbriths half-way on any money proposition. As things are going with me now, I could give Hope a hundred thousand dollars the day she was married. And Jim Langbrith comes of good old stock. I consider his mother the finest lady I ever met, and he's his mother all over again—looks like her, talks

101

like her, *walks* like her. I haven't forgot how she used to come up and nurse my poor wife, and it would be kind of appropriate having the two families brought together again, the same as Langbrith and I used to be in business. Well, now I'd like to get your opinion, doctor. I haven't spoken to my mother-in-law yet, because if the thing doesn't strike you favorably I don't want it to go any further. I want to stop it right here." He lowered his voice from the high note to which it had been climbing back, and looked round him furtively. "You don't think there's any likelihood of that little green fellow coming back? I can get along now with the bones and the skulls, but that dwarf—"

"Have you seen him again?" Anther asked, soberly.

"No, not what you may call *seen*. But I feel as if I might, any minute."

"Well, you know there's no more reality in him than there is in those other things."

"Yes, I understand that."

"But you *will* see him, if you get to letting yourself go."

"Yes," Hawberk assented, with a long breath. "If he wasn't green — kind of mouldy —" He stared, and after a moment he said, "What I want is something that will take me out of myself, good and strong."

Anther was apparently not heeding. He said sharply, "Hawberk! Can you carry your mind back to that old difficulty between you and James's father?"

Hawberk glanced at the doctor, slyly. "What old difficulty?"

"You must know. When you first began to lose sleep, and took up this habit."

The slyness passed into vagueness, then the vagueness gathered itself into a look of fury, which lost itself with the words in which it exploded: "Damn him!"

"Yes. Just what was it?" Anther pursued.

The vagueness came back, and then the slyness. "Why, there wa'n't anything that you may call a difficulty, doctor. All that's past and gone. We both agreed not to say anything more about it. He never did, and I haven't. Ever strike you that that skull of mine—that one I've had so much trouble with—looked like— Well, I've been thinking, since I saw Jim, last night—"

Anther shook his head, kindly. "That's your fancy, Hawberk. But you did feel injured, badly used, at the time, didn't you? Try to think. You know you used to tell me things very different from those you have got to saying since, and I have a reason for wishing to find out the original facts, just now. They may have a bearing on an important matter—important to us both. You put in your invention, didn't you, and then he forced you out?"

Hawberk looked down and passed his hand over his forehead. "There was something like that. But he paid me a good round sum for the invention, and a big bonus for going out, didn't he?"

"You ought to know. Was that really the case, or is it what you've imagined since?"

"Why, I should say it was the case." A fear, the look of a man at some time deeply intimidated, supplanted the slyness in Hawberk's blotched-brown visage. "It's a thing I've agreed not to talk about. He lets me alone because I don't. He's got my promise, and I've got his. If I didn't keep my word, he would be over the wall the first time I fell asleep. You don't catch *me*."

"Come! come!" Anther said, severely. "You mustn't talk that kind of nonsense to me. I tell you, I am quite in earnest, and I would like to know the bottom facts. You needn't be afraid of trusting me with them. You know, as well as I do, the unreality of those troubles of yours. They come from opium, and from nothing else. Now, was there any hold that Langbrith had on you, enabling him to force your consent to going out of business and giving him the entire usufruct of your invention? I want you to answer that fairly and squarely. Was there, or wasn't there?"

Hawberk passed his hand over his tormented brow again. "There was something, doctor. It's strange. It seems as if there was a hold I had on him. But it must have been a hold he had on me. I can't straighten that out. That appears to be the trouble with me. I don't see why I didn't use the hold if I was the one that had it, unless—unless it was something about Mrs. Langbrith. Do you suppose it was? She had been good to my wife; she took care of her in those last days— Oh, my God, how they come back! Doctor, do you wonder I took to it? To get a little sleep! Once I went a fortnight without knowing I had any, if I had any. It was hell. Nothing since from the opium— Ah, I can't think it out!"

"Try," Anther insisted. "It is very essential." He rose from his chair, and began to walk restively up and down the room, while Hawberk lay dreamily staring at the windows, which Anther's person showed itself against, now one and now the other, as he paced to and fro. "I—I don't know but I'm getting it," he began. The doctor's foot struck a plank that gave under it, and the door of the long case fell softly open. Hawberk scrambled to his feet with a shuddering cry, "Oh, for God's sake, what's that?"

"You know—you know well enough!" Anther shouted. "Don't be a fool! It's that old skeleton that you've seen a hundred times. I've had it ever since I've been here, and Hillward before me. Now, do have a little sense. It's a *real* one, and it can't bother you like those fancy ones of yours."

He went up to Hawberk, and put his hands on him to stay his trembling. "What—what does a man want to keep a thing like that around for?" Hawberk faltered out, helpless to take his eyes from the quivering thing that slightly turned as it dangled. "Shut it up!"

Anther obeyed, and Hawberk dropped nervelessly into his chair. "Lord, I don't see how I'm to get home."

The doctor looked at him grimly, then pityingly, then despairingly, as to any hope of further light from him then on the point he wished to clear.

"I'm driving up your way. I'll take you. There's my buggy at the door."

"Oh, thank you, doctor," Hawberk said, and he found strength to follow him out into the hall where his hat and coat hung, and got out of the house first.

XIV

LATE in the afternoon of the following Saturday, Anther stopped his horse in front of the business block, as it called itself, where Judge Garley had his office over the National Bank. It was the only brick building in Saxmills, and the office of the judge was itself approached by an outer stairway of the prevalent wooden construction. The doctor met him on the landing at the moment when he turned from turning the key in his door.

"Going for the day?" he asked, with a disappointment which he could not keep out of either his face or his voice.

"Not if you're coming for it," the judge placidly replied. He turned the key in his door again, and hospitably threw it open. "Walk in."

"Oh, I don't know," the doctor apologized, but the judge took no heed of his apology, except to push him in.

"Sit down," he said, and he reached a book from the top of his roll-top desk. "There's something I think you might find interesting. It's more in your line than mine, and *I've* found it interesting. Well, it's important as a matter of medical jurisprudence, too."

"What is it?" the doctor asked, listlessly turning

the book over in his lap and fluttering the leaves absently.

"Why, it's a study of the criminal settlement on that island off the northeast coast of Japan where the Russians colonize their murderers. As they have no capital punishment, except for political offences, they have to do something with their homicides, and they collect them on that island and keep them there for life. It's very curious, especially in its reversion to some old-fashioned theories —the book, I mean. When I was on the bench— and it has been my experience as a criminal lawyer, too—it seemed to me that very few criminals suffered what we called remorse. They wished to disown their crimes, to keep from realizing that they had committed them, and they wished to get off from the penalty; but I could not make out that they were consumingly sorry for them. This man seems to think differently, and he says some things to make you think he is right. We generally kill off our murderers before they have time to show remorse, but the Russians keep them, in a kind of cold storage, up there in the latitude of Siberia, and they have opportunities of studying effects that we precipitately deny ourselves the knowledge of. The remorse is long in getting to the surface, but, if this man is right, it is always there, and he has heard it comes out about three o'clock in the morning, in the murderers' dormitories, when they wake rested from the fatigue of their hard day's labor, and begin to think. An interesting phase of their remorse is the pity they feel for their victims."

Anther sat with the book fallen shut in his lap, and he did not seem to have been attentive to all that the judge was saying. When Garley stopped, the doctor asked, "What do you think of a man who takes the life of another's soul—destroys his soul? It was a woman's expression."

The judge smiled intelligently. "I should imagine. But I should doubt whether it could be done. Do you want to engage me for the defence?"

"No," said Anther, falling in with his humor, "he's out of danger from the law—unless—unless some law follows up such fellows where they go."

"The old theory was that some law did," the judge suggested.

"Yes, and we can't tell how much truth there was in it. The base of doubt in me is the immunity which wrong-doing seems to have here. But perhaps it's only an appearance."

The judge laughed now. "It serves the purpose of a reality in a great many cases. What scrape do you want me to get you out of?"

The doctor got no further than smiling, though he fell in with the judge's mood, which is the prevailing American mood in the face of any mystery. "Nothing worse than allowing opium to a man who would take it anyway."

"Well, I see that you've decided on your line of defence."

It was a little time before Anther suggested, at an apparent remoteness from the point, "You were never here in Royal Langbrith's day?"

"No, I came here first after my last term in the

Assembly, when he had been dead some time. I believe you and I came here about the same time, didn't we?"

"No, I preceded you several years. And I had known him before I came here. We were in college together?"

"Am I to infer something against him on that account?" the judge inquired, with the jocosity which the doctor had ceased to share, even by so much as a smile.

"He was the devil," Anther said, with a brevity which was of almost a dispassionate effect.

The judge was amused by the succinctness so far as to observe, "In the case of a living person, that is a sort of language which we should consider actionable, I'm afraid."

"The things I know of that man—"

Anther stopped, and sat staring at the judge's law-books where they stood ranged on the shelves before him, showing their red labels on their sheepskin backs with a uniformity in height and shape, broken here and there by cases of pamphlets and documents, and stray pieces of fiction the judge was fond of reading.

"Would fill volumes," the lawyer interpreted, with pleasant interest.

Anther came back to himself with a sharp "Yes!" and Judge Garley went on:

"Well, now, do you know, I'm not surprised, somehow. I've come upon one or two things lately, in a professional way, connected with the deceased, that did not smell as sweet as the conventional memory which seems to have blossomed in the dust all over the place. It is very curious! I have sensed for a good while, by that sixth sense which we haven't got a name for yet, that there was something hushed up in regard to that man."

"There is everything hushed up," Anther nodded, frowningly.

"And you mean that you can tell me—?" The judge checked himself, with a laugh for his weakness.

"Not everything, because I don't know everything; but I know enough."

"Squalid things—the kind we don't like to handle, or pretend we don't?"

"Squalid, and lurid, too. He was the devil."

"There you are, with your actionable language again! It's well for you that our ex-fellow-citizen is out of the way."

"Do you believe," Anther asked, "that one of us can do another a wrong so atrocious as to confound the sufferer's conscience?"

"Cause his brother to offend? Isn't that rather a question for our friend Enderby?"

"Perhaps. But what do *you* think?"

"I should say that it was a theory which a great many people would like to urge for a justification, or at least an explanation of their misdemeanors."

The doctor's tragic humor broke in a joyless laugh. "Oh, of course, you are right. It is astonishing how these old theological cobwebs hang on in corners of the brain. What a comfort you legal minds are! *Advocatus diaboli!*"

"Ah, aren't *you* playing that part now? I should be quite willing to leave our ex-townsman in the enjoyment of his canonization, but you seem to want to reopen the case."

"No." Anther relapsed into his gloom. "It can never be reopened. That is the worst of the evil that lives after men. It intertwines itself with so much of the good in the survivors that you can't strike at it without wounding the best and gentlest of them. But I want to tell you, Garley, about that man— Or, no! Why should I bore you— burden you?"

"Oh, we always like scandal, even concerning the dead. I've allowed that, and I can enjoy yours all the more because I know it isn't idle scandal. Go on, doctor!"

Anther had risen, and he did not sit down at the courteous gesture towards his chair which the judge made.

"Hawberk has got back," he said.

"Ah!" Judge Garley brightened up. "It's he whom you have been allowing the opium? I supposed he always came back permanently cured."

"This time seems to be an exception. He has come back cured of seven-eighths of his ordinary dose, if you can believe him; which you can't. I used to think I could follow his lies, or their probable direction. But I give it up. Beyond having mostly an optimistic character, and being the absolute reverse of the known fact, his mendacity is an ever-new surprise. I give it a harder name than it ought to have. He doesn't mean to deceive, poor soul. It's pure romancing, absolute fiction, but it's no worse. What interested me to-day was the turn which he has taken towards the memory of the man who ruined him. He wanted to persuade me that Royal Langbrith was a fine fellow, with whom he had always been on the best of terms. The fact is—do you want to hear it? Well, I'll tell you anyway—Langbrith did him one of the deadliest and cruelest injuries that ever a man had to bear. You know they were partners in the mills here?"

"I have heard the poetic legend that Hawberk was an ingenious mechanic to whom Langbrith gave a share in the business, and then had to get rid of because he was an opium fiend. Is the legend a little too florid?"

Anther seemed to restrain a burst of fury. When he spoke it was quite pacifically. "You can decide. Hawberk *was* an ingenious mechanic, whose invention put the business on a prosperous basis. He discovered how to make from straw-pulp the light quality of printing-paper which is the specialty of the mills to-day, and which they still have the secret of. Langbrith wanted the whole business. Hawberk had been his partner from the beginning, and he forced Hawberk out under threat of exposing him to his wife, almost maniacally neurotic, in a foolish boy affair with a woman. Hawberk told me, while he could still tell the truth, that there was nothing guilty in the business; but his wife was frantically jealous, and the fact wouldn't have mattered. She would have believed anything against

him, because she must." From his own science
the judge acknowledged with a nod the point which
the doctor made from his. "He brooded upon his
injury night and day, till the night and day were
one, and there was no sleep in them. Then he took
to opium. I prescribed it, as I should have to do
again in a case like his, if we were back where we
were then with soporifics. He could not have taken
chloral. But the opium mastered him, while he was
still hoping for justice from a man who did not
know what justice meant. His opium-eating could
not be kept a secret in a place like this, and Langbrith had it all his own way. The things that *can*
be kept secret are the kind of things *he* did. He
had two wives: one, the woman he threatened Hawberk with, in Boston, and never married, and one,
the Mrs. Langbrith you know here. He went to
town for his debauches of all kinds, and sometimes
when he came home so much of his drink-fury remained that he taunted his wife with the other
woman. He used to strike her—she has told me,
because she had to, and that is how I came to know
the other thing. She told me that, too, but not until it could not be kept from me any longer. What
graves women are for the wickedness of men! I
suppose you know it, in your profession; but in
mine—!"

Anther had apparently come to an end, but he
sank into the chair he had left.

"That does put another complexion on it," the
judge said, sobered in his irony, but ironical still.
"I don't know that I can dispute your professional

superiority as a repository of family mysteries. Your case rather goes beyond any I could boast of, in some features."

"And this," Anther broke out, taking away the handkerchief with which he had been wiping his face, "is the man whom that poor young fool wants to put up a tablet to in the front of the Public Library!"

"I *noticed*," the judge said, "that you seemed to receive the suggestion rather conservatively the other night. I laid it to envy of the deceased."

"Oh, pooh! I know what you mean. I was going to tell you. I do wish to marry her. I don't think she is perfect, and I'm long past the time of marrying for 'love,' as it is called. But to me she is the most sacred of human beings. I have known her from the first days of her hideous marriage, almost from the time when that man took her from her hard work in his mills and made her his slave; for she was that from the beginning."

"Excuse me," the judge interrupted. "I oughtn't to let you go on, if you think I meant to imply what you have inferred. I didn't intend to insinuate that you had the envy of a successor."

"Oh, it doesn't matter. I don't mind *your* knowing what I've told you." Anther stopped there, as if he had lost the thread of his thinking, and the judge made his attempt to restore it to him.

"I had understood that Mrs. Langbrith was the daughter of a minister in the country near here, and was employed in the business in a clerical capacity."

"She wasn't," the doctor said, bluntly. "Her father was a starving saint in the hill parish where she was born; but when Langbrith married her she was a hand in the mills, like forty other girls. It was her inherent dignity that may have given him the notion of something else. Or it may have been *his* dignity. At any rate, he married her and martyred her, even to the blows that fell upon her body as well as her soul. I don't say he never fancied her; and she fancied him, poor soul, as long as he would let her; and when she lost all faith in him she was still his faithful victim. She was so gentle that, though she suffered, she could not resist evil. She was born to keep that commandment. He could outrage her nature, and abuse her to his heart's content, and he could count absolutely upon her silence. He was as safe from her as from the God he found so complaisant to his wickedness."

"Oh, come," the judge remonstrated, ironically still, though he felt the indignant passion that throbbed in Anther's words and respected it; "you mustn't allow yourself to arraign the Deity for His way of doing business. How do you know but our friend is paying his shot now in what is not perhaps 'the easiest room in hell'?"

"Why I have come to you"—Anther made another of his abrupt breaks from the direct line—"is because I want you to advise me what to do. It is all open between her and me, but she has to live in subjection to some one, and she lives in subjection to her son. She has never positively deceived him in regard to his father, but she has never found the time to tell him what sort of man his father was."

"It would have been difficult," the judge owned, somewhat more gravely.

"I have thought the matter over a thousand times, and tried to imagine some moment when she could have spoken to undeceive him, but I never could make it out. All that I could make out was that every moment's delay rendered the truth more impossible." The judge nodded his large head in unconscious assent. "As time went on, the man became a sort of town myth. He grew into the tradition of a conscript father, the founder of our prosperity, the benefactor of the community; and it would have been an insult to the public faith, as well as a terrible ordeal for the boy, if his real character had been proclaimed."

"I see," the judge assented, with a certain pleasure in the perfection of the situation.

"It became a sort of moral necessity," Anther continued, "to leave the past undisturbed, to let the lie remain. The only man who might have unmasked Langbrith living was held from it by the grip Langbrith had of his throat, and Langbrith dead has been safe from him through the optimistic turn his opium craze has taken in the direction of a legend of close friendship between them. Besides, Hawberk's repute as a liar had become so firmly established that his word wouldn't have counted in a place where Langbrith's fair fame is the richest jewel of the local history. There couldn't be a more acceptable, a more entirely popular, thing proposed in Saxmills than his commemoration in the way his son has suggested. It wouldn't cost the town anything, and it would be such a credit to it!"

The doctor laughed for helplessness, and the judge joined in the bitter merriment. "Yes," he assented, with the ponderous movement of his mind which found expression in his formal and weighty diction, whether he joked or whether he adjudicated. "There appears, as you say, doctor, to be a sort of moral necessity to let lying dogs sleep, if we may reverse the axiom, especially when they have slept long. What is the advantage, the better element might ask, of rending the veil of oblivion from errors which Providence has seen fit to leave in the shadow, doubtless for some wise purpose? The morals of the community, they would contend, would be more contaminated by the effluvium from our late fellow-townsman's tomb, if we were to open its ponderous and marble jaws for the purpose of drawing his frailties from their dread abode, than if we were to leave the past undisturbed. They could argue that his success, which is now an example and an incentive to light our young men on the upward way, as long as they suppose it to be founded upon virtue, would be a means of endless corruption if it were known to be the putrescent splendor of his moral rottenness, and would prove an *ignis fatuus* to lure them into the abyss. As far as our community is concerned, doctor, I think you will do better to leave the late Royal Langbrith's memory alone."

"As far as the community is concerned, Garley,"

the doctor returned, hotly, "I think you are perfectly right. But that isn't the point, except for the psychological publicist, if there is such a thing. I am interested solely in the personal view."

"And what is that?"

Anther replied, after a moment of silent chagrin: "I hoped you might have inferred. But it is simply this: Mrs. Langbrith and I both have the belief that our marriage would be abhorrent to her son, not because he has any dislike for me—he is rather fond of me, and I like the boy, when he is off his high horse and isn't patronizing me—"

"He patronizes me, too," the judge observed; the doctor ignored his reflection in proceeding.

"—but because he has this extraordinary infatuation for his father's memory, and would consider his mother's second marriage with any one a desecration not to be voluntarily endured. Simply, she is afraid to bring our wish to his knowledge, and she is afraid to let me do so. I have been almost a father to the boy from his first years; and, under ordinary circumstances, there would be no reason to suppose that our marriage would be distasteful to him. But as it is—"

Anther stopped, and the judge said, with the air of summing-up, "Your conclusion is that the defamation of his father is the only means of—"

"Why do you speak," Anther cried out, "as if his father were an innocent man, and not the wickedest and filthiest scoundrel that ever lived?"

"My dear old friend," said the judge, leaning forward in his rocking-chair and laying his hand on the doctor's arm, "let us be careful not to employ actionable language, even in regard to those who can only cite us to appear before the higher tribunal, which has no jurisdiction in this county, or, so far as I know, in this State. I quite enter into your feelings, and I should be the first to wish you joy of the fulfilment of your hopes, the fruition of your wishes. But you will certainly not further them by adopting anything like a violent line of expression. Now, go on. The boy has returned to Harvard. Have you seen Mrs. Langbrith since?"

"I parted with her in anger Sunday night. But she had tried my patience beyond endurance. She proposed to me, as a way of propitiating James, that—" Anther choked, and the judge had to prompt him:

"She proposed to you—?"

"Well, that I should humor his notion of putting up this memorial to his father; that I should stultify myself, and help to perpetuate the—the—"

"Careful, careful!" his friend suggested.

"Oh, you know what I mean! I don't believe she felt the enormity of it as I did. She couldn't, in that meek forgivingness of hers. But I left her in anger—yes, for the first time; and I don't see my way to making her understand the shame, the iniquity—"

"Really, you ought to have been a doctor of divinity! I think we can leave your reconciliation with her to nature," and the judge finely smiled at the doctor. "But now, in regard to the son's undeception—or shall we say enlightenment?—is it your notion that some third party might undertake the task of accomplishing the end desired?"

"Oh, I don't know what my notion is," Anther replied, rising with a finality which he expressed in superfluously buttoning his coat about him. The day was a warmish day in April, and he might well have found his winter great-coat uncomfortable, even in driving. With the afternoon sun pouring into the thinly shaded windows of the judge's bare office, it was almost a summer heat in which he had been sitting. He added, with a quick sigh, "I didn't know but you would be able to advise me—"

"I will think it over," the judge promised, with bland placidity, and he turned from taking leave of his friend and rearranged some papers on his open desk. "By-the-way," he called after Anther, "I meant to ask you: the brother, who has charge of the business, does he know anything of this double life and character?"

"John Langbrith?"

"Yes. How long has he been in charge?"

"Oh, ever since Langbrith's death. Somebody had to take hold of the business. He was here before that."

"But nothing has ever passed between you and him as to the facts?"

"Not a word. They were not things I could speak of first, and John Langbrith speaks of nothing. I suppose he talks business, but I have no business to talk with him."

"Does Mrs. Langbrith know whether he knows?"

"We have never mentioned the matter, but I

don't believe she does. You know how close he is. He never goes to her except on business, and she has never seen the inside of his house. The mill is his home."

"In his way, he is as successful a secret as his brother?"

"Quite," the doctor said, gloomily.

XV

On his way home to the early tea which Mrs. Burwell's primitive tradition obliged him to accept, in place of anything like a late dinner or later supper, Dr. Anther drove by the Langbrith mansion, and looked hard at it. He turned, when he got by, drove back, stopped his buggy at the gate, and hurried up the brick walk to the door. It was opened, before he could ring, by Mrs. Langbrith. "Both the girls are out," she partially explained, and she could have said further that the middle-aged serving-women, who were still girls to her, had not outlived their youthful passion for mingling with the crowds which thronged the long main street of Saxmills on pay-day, and that she had yielded to it for the sake of the pleasure which the fine weather would add to their outing. But he paid no heed to her opening words, and she did not go on.

"Amelia," he said, with the fervid rashness that was natural to him, "I want to beg your pardon for the way I left you last night."

"Oh!" she murmured, so deeply that the murmur was almost a sob.

Then these two elderly people did by one impulse what they had never done before. In the dim hall, beyond which Anther had not tried to penetrate, they cast themselves into each other's arms, and he kissed one cheek of hers, while she buried the other in his neck, and smoothed her silvered brown hair, and kept saying softly, "Poor girl, poor girl, poor girl!"

He kissed her cheek again, and then he walked slowly and thoughtfully down to the gate, and got into his buggy and let his horse take its own gait and course. Not only a tender patience with her swelled Anther's heart, but an unwonted tolerance for young Langbrith also found room for itself there. What wonder that the boy was reverent of his father's memory, since he knew no evil of him? Was it for this he must be called fool and despised for an ass? Anther saw that there were yet many steps to be taken with regard both to him and his mother, and that they could not be separated in relation to himself. He softened more and more towards the whole situation, and momently the thought of the weakness he had surprised in her consecrated and endeared her to him.

He drove along the village street with his figure stooped well towards the dash-board, when his ears were saluted with a succession of girlish trebles.

"How do, doctor!"

"How do, doctor!"

"How do, Dr. Anther!"

He looked up blankly, and presently realized that he saw Hope Hawberk, Jessamy Colebridge, and Susie Johns, walking, with arms more or less intertwined, along the pavement which he was closely skirting by the erratic preference of his horse.

They smiled brightly upon his daze, and nodded gayly to him, hanging over one another and laughing at him over their shoulders when they got by. He gathered himself together to call back to them, "Oh, how do you do!" and the charm of their differing prettiness very sweetly possessed him. They were like his own children to him, in his long intimate acquaintance with their ailments as a physician, and with their accomplishments as chairman of the school board. Their young voices, and their arch, familiar, trustful tones made the blood play warmly about his heart, and he let his horse take him home to supper in a mood which he could not have imagined of himself when he parted such a little while ago from Judge Garley.

The girls walked on down the street towards the denser part of the town, chattering, singing snatches of song, humming and laughing, leaning over to mock one another, and then straining outward or forward in their fun. They sobered as they got more into the crowds thronging the sidewalks, till they distinguished themselves from the mill-girls by a demure state, which could not leave one in doubt of their quality as village girls who did not work in the mills. Mill-hands of both sexes were exuberantly filling the street, after their release from the week's work, in a tumult of shopping, of carrying on, of courting, which would last far into the night. The young men stood at the corners or lounged along the curbstones, smoking, and challenging the girls to a stand which here and there stopped the way with giggling and slanging and tussling groups;

the girls, when they did not stop, tossed chaff and
sauce at the young men over their shoulders and
tempted them to pursuit, as they passed chewing
gum. But neither the young men nor the girls
molested the three friends, who had now separated,
and were pushing sinuously through the open spaces
towards the post-office. The mill-hands knew who
each of them was, and how they were nearly always
together; some had been in school with them,
or in Sunday-school, and these exchanged nods
with them; others who were strangers to them
looked inimically after them, as representatives of
class.

The three were not equally friends, though they
were all friends. Hope Hawberk was chief among
them, and Susie Johns was next her in the understanding that Jessamy Colebridge was capable of
being silly at moments when the others would rather
have died. Without being untrue to her, they sometimes laughed a little at her; but that did not keep
either of them from laughing a little with her at
something queer in the other. Susie and Jessamy
both knew about Hope's father, but her grandmother was of a family which no one in Saxmills
could look down on. Her grandfather had been
Squire Southfield, once the chief lawyer of the place,
and he had been in Congress; though that was a
long time ago, and her mother had certainly married beneath her in taking Hope's father. He was
then a skilful young mechanic, but it quite passed
the knowledge of Hope's friends that he had been a
very fascinating fellow, whom such a girl as Hope's

mother could not resist. Hope was like him in the dark coloring of her beauty, her dusky hair, and her black eyes; but there was a passionate irregularity of her mouth when she smiled which was the trace of her mother's stormy temperament. She had really more of her father's amiability, which, to the strict New England sense, erred almost to the guilt of easy-goingness. His dreams had not begun with opium. There were psychologists among his critics who regarded the opium as the logical consequence of his dreams, and who, if they had been asked in time, could have prophesied from the first all that he had come to since.

Neither of the three girls expected a letter, but when it seemed that there really was a letter for Susie Johns, Jessamy confessed her own disappointment with a quick "Oh, dear!" in taking the letter from the girl clerk behind the boxes, who severely announced, "Ain't nothing for you or Hope." But Hope, if she had a disappointment, hid it under a laugh.

She caught the letter from Susie's lax hand, and said, "Let me read it for you, Susie dear," and Susie wrinkled her nose, and said, "Well, you may." But Hope contented herself with looking at the post-mark.

Jessamy joined her in the inspection, and it was she who proclaimed their joint discovery. "It's from Boston! Why, Susie Johns, who's been writing to you from Boston? Oh, I'll bet it's Mr. Falk."

It appeared that the letter was really from Mr. Falk, but not till the girls had left the anteroom of the post-office and made their way back homeward on the up-hill street leading out of the business thoroughfare. Then, when they could have the whole sidewalk to themselves again, each of the others passed a hand through Susie's arms and prepared herself to help her make out any hard words, leaning forward in readiness. Jessamy kept babbling as Susie read her letter silently through, and by the time she reached the end Jessamy was offering the twentieth variant of her wonder: "What in the world is he writing to you about?"

"Oh, it's just manners," Susie responded serenely. "I suppose he thought he ought to write and say something pleasant about his visit here."

"Is that all?" Jessamy innocently protested, and this made Hope laugh.

"What else did you expect there would be?" Susie folded the letter up and put it back in the envelope.

"Well, I don't know. He might have sent some message!"

"He did. He said 'give his regards to all inquiring friends.'"

"Oh, that sounds nice. It's just what *we* say—village people. But I believe Mr. Falk isn't from a very large town. Only you always think students must be like city folks. Dear, I wish I had a letter."

"Well," Hope said, "I'll ask Harry Matthewson to write you one."

"No, you mustn't, Hope. Will you, really?"

Susie squealed, "Jessamy Colebridge, you certainly are almost a goose"; and Hope said, "Well, I won't if you don't want me to."

They had come to Jessamy's gate, and Hope pushed her arm through Susie's and ran her on, while Jessamy stood looking in rueful puzzle after them.

"Jessamy *is* such a simpleton. I should think she was a child of ten yet." Hope put her face down on Susie's shoulder and laughed, and when she lifted it Susie put her face down on Hope's shoulder and laughed. Then Susie offered to let Hope read Falk's letter; but Hope had never shown her the letter which she had got from Langbrith the Monday before.

XVI

Beyond the village, the little lake from which the
mills drew their power had been clear of ice for weeks,
but its waters had kept the look of winter. The
logs weltering at the gates where the current which
was to grind them into pulp left the lake, dipped
and lifted with a cold, wet gleam as they pushed at
the pales on the pull of the stream. A day came
when the whole aspect of the landscape changed.
No leaf had started, and scarcely a bud had swelled
on the water-elms that showed their black trunks
and boughs amid the green gloom of the pines
and spruces overhanging the shores, and the white
nakedness of the birches had not yet clothed itself,
except for a thin veil of catkins. But the water
had taken a warmth of tone from the sky, which
was of a deep blue, heaped with milky clouds roughed to a superficial dusk by the southern wind. Blue-
birds rose and sank with the rhythm of their queru-
lous notes in their short flights about the farmsteads
and village houses. The robins in the chilly morn-
ings and evenings shouted from the door - yard
trees. Ragged lines of blackbirds drifted with a
glassy clatter over the woods and rested in their
tops, or slanted towards the water, where they
showed their iridescent splendors, as they strutted

up and down on the logs and parleyed harshly together.

Hawberk sat tilted in his chair against the southern house-wall where the sun struck into the garden, and listened with a dim smile to their clatter, coming over to him through a cleft of the land that let the lake shine through below the hill. He had begun the joyful day with half a gill of laudanum, and he was feeling the primary effect of the drug in the delicious lassitude he won from it at continually increasing cost. He was smiling, not only at the noise of the blackbirds, but at the comfort of the cat, which had found the stone warm at the base of the sundial in the walk of the little garden, and lay coiled there. He liked the look of the dishevelled beds, where the dry litter of the last summer's stalks and stems was mixed with the tawny blades of the grassy borders, and he liked the softly waving plumes of the pines which beckoned to him from the brow of the hill behind the little dwelling. He heard, with the same sensuous pleasure, the jar of the mills below the street on which the house fronted, and he vaguely recalled the relation his life once had to that busy sound, now no more to him than the idle sound of the wind in the pine tops with which it was effectively one. Exquisite thrills passed through his relaxing nerves, and the twitching of his muscles was divinely voluptuous. Then, suddenly, he was in that pit again, out of which he had slowly fought his way at the Retreat, but which he knew he must now sink back into day by day. The green dwarf was there as he had not been for a long time, and was at his work of slowly filling in the sides of the pit, making it smaller and smaller, and arabesquing its surfaces with patterns of men's bones. He choked in the thickening air and dug his way upward with his hands, toiling for months, for years, for ages; but the pit was always filled in again, and its roof and sides faced with those hideous arabesques. After centuries, he saw the light break through from above; then the dwarf came slowly overhead, and covered him in again and shut out the light. The groans of his torment ascended continually; when the dwarf extinguished the last gleam, the horror was such that it burst into a scream of despair—a cry of agony so sharp that it cut his dream asunder, and he woke with cold sweat, and saw the cat dozing at the base of the dial.

"Father, father!" the voice of Hope called, while she caught his reeking hand in hers.

He tilted forward out of his chair, trembled to his feet, and stared around, gasping.

"Oh, Hope, child, don't let me sleep, don't ever let me sleep again. How long have I been here?"

"Only while I could go in and get my hat and a book to read to you. Grandma wanted me a minute."

"It seemed eternity. Don't let me sleep again. I'm all right if I don't sleep. Promise me that."

"Well, I won't, father. But come now—or aren't you able to go up the hill with me?" He had sunk back into his chair, and she kissed his forehead, blotched from the opium, with its sunken eyes beneath it, and the red scars seaming his cheeks, from which a sickening odor came. "But must you?—*must* you?"

"Yes, yes, I must. Don't talk to me that way. I must, I tell you. If I had a little, now! Where is it?"

"In your room. I'll get it, if you say so—"

"Well, get it then, quick, quick! I don't want to sleep again."

"Don't be afraid. I'll be back in a second."

She vanished, and reappeared with a bottle in her hand which she put into his shaking hold.

He pushed it to his lips without looking at it. When he had drained it he glanced at the empty bottle. "Was that all?"

"Yes, every bit. But I can get some more this afternoon if you want it."

"Of course I want it; it puts life into me. Ah!" He drew a long breath and stretched himself. "That's something like. Now come on." He laid his shaking hand on her arm, and they began to climb the hill together on the path that found its way upward by little juts of the ledge, and little turns round them, and over the rough surfaces where the thin soil left the rock bare. "It's astonishing what it does for a man. It's all that keeps me up, in these enterprises. But don't *you* ever touch it, Hope. It's the best of servants, but the worst of masters. If I didn't know how to control it so well, it would play the mischief with me."

Hope said, with the lightness which all the horror of the situation could not sadden in her, "And even *you* don't seem to have the upperhand *always*, father."

Hawberk laughed in sympathy with her lightness. "That's a fact, Hope. But it's very seldom. The great thing is to know when to pull up. I'm all right as long as I'm awake, and there's nothing like it to keep you awake. You've got to use it regularly if you want to get the good of it."

"Well, you've wanted to get the good of it about two hours too soon to-day, father," she said, with caressing mockery.

"Why, what time is it?"

"About eleven."

"Lord, I thought it was after dinner, and I'd gone by my time. You oughtn't to have given it all to me, Hope. I don't know what I shall do now till night."

"I'll get some more for you from Dr. Anther. He wanted you to have it."

"I don't know about that. I believe he wants to keep it away from me, though he knows it's the only thing that will carry me through this pinch of work. I want you to go right after dinner for it—before he starts on his visits."

"I will, I will, father."

"It's the only thing that will keep me awake, and as long as I don't sleep I'm all right."

"Well, I should think you would find it pretty hard to manage without any sleep at all," Hope said, always in the same drolling fashion. "Why don't you try to stop it altogether?"

"That's just what I'm going to do when I get

through this pinch. I've talked it all over with Dr. Anther. We've got the whole thing mapped out, down to the last dot."

They had reached the top of the hill in their talk, which had had as much silence as parlance in it.

Hawberk let go the arm to which he had been clinging less and less dependently, and straightened his bent, wasted frame.

"Fine! fine!" he said, looking dimly out of the caverns under his brows at the prospect. "I think I shall put the house right here. You know I've bought this hill, Hope?"

"No, I didn't, father. But I'm not the least surprised to hear it. You keep buying all sorts of things." She had settled herself on the warm, brown needles under the pine where he stood; and, as she spoke, she pulled her skirt closely about her knees and folded it under them. He looked down into her face, and they both laughed.

"But this is a fact, Hope. That last little thing of mine is doing so well in the hands of those people at Boston that I've decided to build here. We haven't passed the papers yet, but I've got old Arlingham's agreement to sell. Drew it up yesterday before Judge Garley, and left it with him. I'm going to have an architect make the plans. It's to be for you, Hope."

"Me? Oh my! I like the little old place at the foot of the hill well enough."

"It's well enough for your grandmother and me, but I want you to have a decent place when—"

"Well, well! That's all right, father; and I'm

ever so much obliged. But you better sit down and have a rest before you begin building." She kept the same joking tone, but there was a sort of nervousness in the haste with which she cut him off from the topic, and hastened to say, "I'll read to you now."

Hawberk obeyed, and leaned his bared head against the trunk of the pine at whose foot he sank; his eyes closed, and he instantly started forward, with a shudder and a cry of "Ugh!"

She closed on her thumb the book which she had just opened, and asked, gravely, "Was it the green one?"

"It's always the green one, now," he lamented.

"Well, then, I'll tell you what, father: you're getting pretty bad again."

"No, no! I'm all right—or I shall be, if I can keep awake. I guess you better talk to me, Hope. Better not read. Seems to set me off at once. You'd just as lief talk, wouldn't you?"

"Oh yes. It doesn't matter to me. You'll do the talking, anyway."

Hawberk laughed. "I guess that's about so, Hope. The reason I want you to have this place here is because Langbrith and I used to talk about building here together. We used to be great cronies, Royal Langbrith and I did, and it seems quite appropriate—"

"Now, look here, father," the girl broke in, "you're getting on to forbidden ground. You may choose any other subject to talk about, and I'm with you, but I can't follow you there."

136

"Oh, all right, I wasn't going to. But now let me tell you the kind of a house I've got in my head for this place. Of course, some of these pines will have to come down." He got up, and began to walk about and take in the shape of the ground, and pace off certain measurements, and look up at the different trees. "But I shall leave a row of them in front, and a lot off to the side, here." He gestured towards the right, as he came back, and sat down again. "But all back of here the trees have got to go. I want to have you a good big garden behind the house."

"Well, I'm almost sorry for that," Hope humored his fancy. "I believe I'd rather have the pines than the garden. They do smell so nice, with this sun on them."

"That's a fact," her father assented, sniffing the balsamic odors that the heat drew from the boughs softly stirring themselves in the wind. "Well, I'll leave as many as I can, Hope." He broke off with, "What sort of young fellow is that one who was up here at Easter, with James?"

"He's pretty nice, I believe. What makes you ask?" Her own question had something of the anxiety in it which marked her escape from his approaches to the forbidden topic of Langbrith.

"Oh, nothing. They tell me he's something of a draughtsman—kind of artist."

"Yes, I told you that. What of it?"

"Nothing. But I've thought some of employing him to illustrate the advertisements of that last little thing of mine. Those people down at Boston

137

are going to have it written up in great shape for the back part of the magazines, and I want to have pictures. Suppose he could do them?"

"Yes, I should think so. But now, look here, father: you mustn't go talking this around."

"No, no! I just mentioned it to Dr. Anther the other day. He thinks very well of it."

"Did you say anything about James, to him?"

"No, no, no! Not a word."

"Nor to anybody else?"

"Why, I haven't been home long enough to see anybody else."

Hope left that subject. "Well, now, I'll tell you what, father. I think after you get through this pinch, as you call it, you had better talk with Dr. Anther about leaving off, gradually."

"Why, that's exactly what we did talk about the last time I saw him. We've fixed up a splendid plan. The doctor's all right. I told him what I thought the weak points at the Retreat were, and he agreed with me right along. He's going to study into my case. It's peculiar. I've kept it up so long, and yet there hasn't been a day when I couldn't have left it off. My idea is to stop the thing short off. No dilly-dallying."

Hawberk's words expressed an energy which his weak tones and his stumbling gait in his restless movement to and fro as he talked altogether belied. Hope sat watching him, with a face which her mocking words in turn belied when she spoke.

"Do you think you can manage to catch a little waking spell every now and then till I've been to the doctor's? I don't want that green one round when I'm gone, even if he isn't real."

Hawberk laughed joylessly. "He's got to stay away till night, now, anyway. I can manage him. Don't you be afraid. I'll get your grandmother to give me a good strong cup of coffee at dinner, and that will help to keep him down."

"Well, shall I read now?"

"Yes, read away. I'll keep moving; or if I get to dozing when I stop to rest, poke me with this stick."

He gave her a fallen bough which he stripped of its dead needles and broke to a stout club, and she took it in the drolling humor which formed the atmosphere of their companionship.

There was enough of this feeling in her face and voice to make Anther pause a moment when she asked him, a few hours later, "Doctor, can't something be done about father?" She sat with the stout bottle of laudanum which Anther had given her in her hand, and tilted it back and forth on her knee.

"How do you mean?" he finally asked.

"Well, to make him stop it."

The doctor rose and closed his door, and then sat down again and kept his eyes absently on her smiling face, as if his mind were at work beneath its surface, seeking the measure of her portion in the suffering to which her young life was helplessly related. He was not likely to exaggerate her sympathetic suffering. He had seen how the young life is always defended from the worst misery of the old; how from their common source it flows on in

139

the same channel, and takes no tint or taint from the concurrent stream, but keeps itself pure and glad side by side with the darkest anguish.

"Do you know how much he's taking now?"

"I guess he's got back to nearly the old quantity."

Anther waited again before he spoke. "I didn't expect it so soon after he had got home."

"I don't think the Retreat did him much good. But I believe *you* could, Dr. Anther."

"I don't know, my dear! Does he believe it?"

"Oh, he believes in you; and I know he would like to make an effort to stop it. I know he'd help you. I don't know what he's going to do. He has got to sleep, of course, but the minute he goes off he begins dreaming, and that green one comes, he says, and tries to wall him in. It's pretty awful." She laughed in a queer way, and then the tears burst from her eyes. "You must think I'm a strange person, to laugh at such things."

"No, no," the doctor said, tenderly. "I understand, Hope."

"I suppose it's my being used to it all my life that I don't realize it as some others would. And then father is so funny when he tells about it, and acts it out, as he does. I suppose I'm like him. He knows it's nothing, as well as you do. But it's real while it lasts."

"Yes," Anther said. "But you're right not to distress yourself about it, Hope. That wouldn't do any good, and you can help your father best as you are."

"Well, I am afraid I am of a light nature. Grandma says so. Now and then it all comes to me, what he goes through, and then"—she quivered on the verge of a sob, but controlled herself and said, "Well, I didn't make myself; and I haven't got myself to blame for ever forgetting him, anyway."

"I know that, my dear." Anther sat thinking, till Hope recalled herself to him.

"Don't you believe it's worth while to try again, doctor?"

"Yes, indeed! We must never give up trying."

Anther rose again, and opened the silk-lined glass doors which shut in the shelves where he kept his office-supply of drugs, and began mixing a bottle from various bottles before him. He shook the mixture vigorously, with his thumb over the mouth of the bottle, and then corked it, made a little pencil-mark on the top of the cork, and gave the bottle to Hope. It was quite like the bottle of laudanum, in size and shape. "There!" he said. "I've mark-ed the cork so that you'll know it, and I want you to keep it where you can substitute if for the lauda-num every other time. Understand?"

"Yes, I understand, but—"

"It won't hurt him if he gets the laudanum bottle, now and then, instead of this; it may even help to tide him over a bad place. But try to make the alternations regular. Gradually—"

"Yes, but hadn't he better break it off altogether—at once?"

The doctor shook his head. "It might do in some cases, but it won't do in his." At something insist-ent in the girl's face he said: "You want a reason? Well, because we've tried it once. It was a good while ago, when you were little, and before you were old enough to know anything about it. We agreed to stop it short off. We agreed with Wason, the apothecary, he was then—young Wason's father— that he wasn't to let your father have anything without my orders on any conditions whatever. I took his laudanum away, and the third night he came to me half-dressed, through the blinding snow, and woke me, and made me give him the laudanum. I have always been humbly thankful that I had the sense to do it, and I have never tried to stop him short off since. I tell you this, for I don't want you to let him tempt you into any experiment like that. He is quite likely to smash his laudanum and try to go it on the other alone."

"I know it!" Hope smiled in recognition of her father's optimism. "He does feel so sure of himself when he makes his good resolutions!"

She rose, with a large bottle in either hand, and the doctor, seeing how she was cumbered, said: "I'm going up your way. Get into the buggy with me, and let me take you home. Nobody else seems coming to-day."

When she was tucked in beside him, he let the old horse jog at will in the direction he had given, and resumed the talk broken off in the office. "Does he take to the same hopeful view of things generally as ever?"

"Well, whenever he can get away from the green dwarf, he does," the girl said. "You know," she smiled across her shoulder into the doctor's face, "he has bought the hill back of our house?"

"I think he mentioned it," the doctor returned, with the same quality of smile.

"Yes, he's going to build for me there. Nothing can stop him. Doctor," she went on with a note of tragical imploring which had not got into anything else she had said of her father, "did he speak to you about—about—James Langbrith?"

She gasped out the name, and nervously put her hand on the doctor's, pinching the buckskin of his glove between her little thumb and forefinger. "Because there isn't—there isn't— Oh, it would kill me if I thought he was talking to people!"

"Oh, poor thing!" said the doctor. "Don't worry! He did speak to me, but, of course, I understood."

"Oh, I don't mind his speaking to *you*," she said. "You're like one of the family; but—but—"

"Well, you needn't be afraid. You know he sees almost nobody out of your own house but me; I cautioned him against talking of that matter, and he usually regards what I say."

"I suppose it's just the dreadfulness of it that scares me. But it would be more than I could bear. Will you speak to him again, doctor?"

"Yes, yes, I will, my dear. Don't you worry!" Anther turned his face away, and smiled to realize that the girl who could keep her courage in the face of misery like her father's should lose all her strength at the thought of having her name coupled with the name of the young man who loved her, and made the talk of the village. But that was youth, and that was life. "Don't you be troubled!" he said, looking at her again. "Nobody would mind what he said."

"Is that much comfort?" she asked.

"It's the most there is," he answered. They drove along in silence broken by the rattling of the loosened nuts in the framework of the old buggy, and the dull clump-clump of the horse's hoofs on the road. Suddenly, as if at the end of a sharp decision, he asked, "Hope, does your father ever speak of James's father?"

"Why, yes, he always says what friends they used to be—cronies. He says he was the best friend he ever had. He was, wasn't he?"

"Oh yes—yes," Anther said in a lie that sickened him; but he had brought the necessity of it upon himself, and he could only hang his averted head in merited shame. "I didn't know but sometimes he took the other turn. You know," he went lying on, "how his mind works by contraries."

"Oh yes, I know that," Hope said, and she did not reason to the corollary, in her concern with the more personal fact. "But that wouldn't help if he got people gossiping about me."

It came to Anther again, as it had come before, that each generation exists to itself, and is so full of its own events that those of the past cannot be livingly transmitted to it; that it divinely refuses the burden which elder sins or sorrows would lay upon it, and that it must do this perhaps as a condition of bearing its own. He idly flicked the road with the lash of the whip which he so seldom laid upon the back of his lazy old horse.

XVII

A LETTER for Hope came from Langbrith the day
after he went back to Cambridge, and letters had
come from him at frequent, irregular intervals since.
They were nearly all of the same tenor, growing
more urgent and impatient in their protest of his
love for her, and in his demand for some answer
more definite than she had been willing to give. She
had gone no further than to say, "I do not know
whether I care for you or not, in the way you mean.
I should not think our being children together had
anything to do with it. If it had, I ought to hate
you, because you always used to try to domineer
over me. If it is any comfort to you, I will say
that I do not hate you, but that is the most I can
say now. As for promising anything, that is ridiculous as long as I am not certain. I am going to
keep myself as free as the air, so that if any one
comes along that I like better I shall not be bound
to refuse him. But there are such droves of young
men passing through Saxmills all the time, I may
not be able to choose. If anything can make me
choose somebody else, it will be asking so much to
choose you. I don't like to be followed up."

Langbrith tried to read the meaning into her letters which he could so little read out of them. But

when it came to this last declaration of hers he
thought it best to forbear, and in his answer he held
his hand altogether. He did not recur to anything
she had said, but made his letter, not without resentment, about Falk and their contributions to
Caricature, and about some teas and dances which
he had been going to in Boston. He wished to
philosophize these social facts, and contrast the
manners and customs of Saxmills with those of the
town. It was his conclusion that, with some superficial advantages, the city was not politer than the
village. "The society buds here have a rudeness
which strikes me as worse than the freedom among
our village girls, which would shock them. People
talk of the decay of social life in the country; but I
shall be very well satisfied to settle down at Saxmills, when I have got all my tools, and go to work
there for life. By-the-way, I hope you will be interested to know that I have been talking with that
young sculptor here whom I told you about, and he
has taken my idea of a medallion of my father in
a very intelligent way. He is a great worshipper
of Saint-Gaudens, and he is quite with me in not
wanting to do anything round or oval. He thinks of
an oblong, with the greatest length horizontal, for a
head of my father; in the upper left-hand corner, an
inscription of three or four lines, with dates and the
name, and in the right corner, a relief of the mills as
they looked when my father first took hold of the
business. He did want to have him holding a relief
of them in his right hand, as people are shown holding cities and temples in some of the old sculptures,

but I am afraid this would not be understood, and I do not want to have anything that could detract from the serious feeling which the tablet ought to inspire. I wish you would think it over and tell me how the notion strikes you. Don't talk with any one else. I want your opinion alone. How would it do to have the dedication on Decoration Day?"

Hope wrote back a scoffing answer, so far as concerned the appeal for her judgment in such a matter, but she freely gave it against the archaic treatment. She said it would look funny. As to the best time for the ceremony of dedicating the tablet, she refused to say anything whatever. But she did say that it seemed to her Decoration Day belonged to the few old soldiers who were left and their families, and it ought to be left to them. It appeared that this notion struck Langbrith as of the most immediate importance. He did not wait to write an answer; he telegraphed: "Thanks about Decoration Day. Perfectly right. Would be ridiculous."

The telegram was brought to Hope while she sat trying to talk her father out of a plan he had for taking Dr. Anther's prescription only half as often as directed. His reason was that he had proved its efficacy so thoroughly that there was no hurry about his cure. He was satisfied now that he could drop the opium habit whenever he liked; but, at present, just while he was working at a new invention in his mind, he needed the tonic and strengthening effect of the laudanum. Hope argued the question with him half jocosely, as she treated all

147

the phases of their common tragedy, and prevailed with him to continue the doctor's treatment to the end of the week. "If you stop it now," she urged, "you'll have that green dwarf back the first time you drop asleep, and I can't stand him. He's made more trouble for this family—!"

The grandmother, a fierce little spectre of a woman, with burning blue eyes and a whorl of white hair crowning her wrinkled face, could not make the father and the daughter out. She kept the house-keeping fast in the strong, shrivelled hands into which Hope's dying mother's hands had let it fall, but she did not meddle with the girl and her father except in the way of censure and prophecy of doom. "If I had my say, I should fill that laudanum bottle up with good strong, black coffee, and not let him have anything but the coffee and the medicine."

"Then you'd have him tearing the roof off. Father would know the difference between coffee and laudanum the first sip," Hope said.

"And is it a daughter's place to give her father poison?"

"It seems to be *this* daughter's place, grandma. Besides, it isn't poison for him, and it's Dr. Anther's orders."

"Oh, a great doctor! I tell you, child," and the old woman flared her fierce visage close in the girl's face, "it won't be the doctor that will have to answer for this."

"Well, I hope nobody will. There must be a great deal of harm in the world that nobody in particular has to answer for."

"Do you mean to tell me," the old woman demanded, "that all the sin doesn't come from sinners?"

"Now, grandmother, you know I don't understand about those things, and I never did, even when I was little and expected to. You'd better ask the people at evening meeting some time. I can't tell you. All that I know is that I'm going to follow the doctor's directions in spite of father and you, both, and I'm not going to make it all medicine or all laudanum to please either of you. What is it, father?"

Hawberk had gone down to the side gate at the first menace of dispute, and left Hope and her grandmother to contend over him, while he remained beyond the hearing of the censure which the old woman could always make him feel that he merited, though he had his theories that he was the helpless prey of his evils. Hanging over the gate in his nerveless fashion, he was approached by the boy from the telegraph office, who preferred climbing the hill on a bicycle to bringing a message less laboriously on foot. At sight of him the old woman quenched her flaring presence in the dark of in-doors. She was afraid the boy had heard her lifted voice, and Hope sauntered across the grass while the boy was taking the despatch out of the inside of his cap.

Hawberk looked at the address, and then handed it up over his shoulder to her. "Why, who in the world," she wondered, "has been sending *me* a telegram? Dear, I wish they wouldn't, whoever it is," she said in a laughing panic. And then, having

149

read it and frowned darkly at it and puzzled over it in a second reading, she started back to the house with the laugh, but none of the panic, and the proclamation, "Well, certainly, he is the greatest—"

"Any answer?" the boy demanded, as sternly as a boy could in supporting himself on his stationary wheel by holding to a picket of the gate.

"No, of course not," Hope called back, and she added, in a lower voice, "Goose!" which, if it was meant for the boy, did not reach him in the swift scorch on which he had instantly started down the hill, in compensation for his difficult climb.

Her grandmother, lurking in the shadow of the cramped entry, tried to stop the girl in her flight up the sharply cornering stairs to her room in the half-story. "What is it, Hope?"

The girl called down from above, "Just some nonsense from James Langbrith," and, with the telegram flattened and reperused on her table before her, she began to write.

"I have just received your despatch. At first I thought it must be somebody dying, or telling me that I had been left a fortune; but I decided against that before I opened it. Of course, I am proud to think my opinion is so important that it has to be acknowledged by telegraph. But I guess you had better wait and write the next time. I was not very likely to run off and see the Selectmen and have a town-meeting called before I could hear from you by mail. I hope you will not be disappointed if I don't telegraph back. But if everything you have anything to do with is so important, perhaps you will be. I don't know what that new telegraph-girl at the depot will think. She must be trying to puzzle it out by a cipher code and racking her brains over it. Why *did* you send it? Did you think what you had suggested was so very silly that you could not bear to let it go another night before taking it back?"

After venting the agitation of her fluttered nerves in these railleries, she went on to make Langbrith what amends for them she could by writing a longer and friendlier letter than usual; and, when she had finished it, she told her grandmother she was going to the post-office, and perhaps she would stop to see Susie Johns on the way, but she would be back again soon. She tilted down the long hill-side street, and her face was as gay with the fun reverberating in her mind from her letter as if she had left nothing but a sunny serenity in the house behind her, where her father was fighting away from the horrors of his dream, and her grandmother was gloomily exulting in the doom that must follow his ill-doing, as if for the reward of her well-doing. While Hope was with them, she felt the oppression of their unhappiness; but out of their presence, it existed for her only as something inevitable, which she must not take any more seriously than James Langbrith's self-importance. The unhappiness made her laugh sometimes, as Langbrith's pomposity did, or the thought of his clumsy truth and the humble pride with which he owned himself wrong in his absurdities.

"What long steps you take!" a voice called after her at a corner she was passing, and she whirled her

face over her shoulder to see Mrs. Enderby hurrying
to join her. "Hope," the rector's wife said, breath-
lessly, "you're the brightest and blithest thing in
this town."

"Am I, Mrs. Enderby?" the girl laughed, slowing
her pace for the friendly lady.

"Don't you know it? Or perhaps you don't,
and that's the reason why you can keep it up. Don't
try to realize it, child. How are you all at home
this lovely morning?"

"Oh, we're always well, Mrs. Enderby. That is—"

She stopped, and Mrs. Enderby went on for her.
"I'm not going to make you conscious, and you
mustn't let me, but just to see that face of yours is
inspiration. Were you always so?"

"Why, I don't know what kind of 'so' you mean.
I suppose I'm pretty well all the time, and that
makes a difference."

"And I'm not going to tell you that you're pretty
without the well, for that never makes any part of
the difference. But, Hope, you *are* pretty, whether
you know it or not."

"Well," the girl drolled, "I don't know as I could
do anything about it if I did."

"No, and that's what makes me feel so safe in
praising you. I know it won't spoil you. When
you came rushing along past the corner, you made
me think of some tall flower sloping in the wind. I
wish you would tell me just what flower you made
me think of! If there was some kind of black iris!
Well, I must try to find out."

They laughed together, and Hope said, "If I

knew, I might think you wanted to flatter me, Mrs. Enderby."

"No, I'm not flattering you. If I told you what I thought of you that night at Mrs. Langbrith's, you might suppose I was. I couldn't keep my eyes off you. And other people couldn't. I dare say you didn't know it?"

"If I did, I must have forgotten it by this time; it was such a long while ago."

"Hope, you are not only the gayest and prettiest girl here, but you are the wittiest."

"Well, now, I *know* you're not flattering me. It's no more than my just dues to have you say that."

"Oh, I'm only repeating what I hear other people say. I wonder," Mrs. Enderby went on, as if to the very next thing, "whether Mr. Langbrith spoke to you about a great scheme that he has in mind?"

Mrs. Enderby was launched, and nothing in her own nature or the situation could keep her from sailing to her destination. As a Boston woman valiantly and loyally following her husband, not only from the Unitarian cult in which they were both born into the church on whose ritualistic heights the rector had planted his banner, but also from the many lively interests of her native city into the social desolation of Saxmills, she realized from time to time that she owed herself all the amends within her reach, and she was not one to be guilty of the injustice of withholding them. She had been charmed with Hope from the first, and when she perceived, as she did very early in the history of her establishment in Saxmills, what this poor, pretty, happy, tragical creature obviously was to the young owner of the local industry and prosperity, the mother-heart of her childlessness bowed itself upon them both, and held her in the hope of at least so revealing them to each other that they need not err as to their mutual meaning. The affair satisfied the most recondite demands of her soul by its romantic properties; and that disparity in the worldly fortunes of the pair did not affect her with a sense of unfitness, as it might have done if they had been Bostonians. They were both natives of a place that, without any sort of social traditions, had grown from a small village under the magic of the elder Langbrith's enterprise into the busy little town she knew; and the picturesque legend of Langbrith's forbearance with the infirmity of Hope's father until he could forbear no longer, touched the fancy of Mrs. Enderby as the material of a peculiar tie between the young people. Something better than her fancy was pleased with the notion of the father's reconciliation in their children.

"About what scheme?" Hope asked, with the inevitable hypocrisy.

"He was speaking of it to the gentlemen after the ladies left the table that night, and Dr. Enderby mentioned it to me. Why! I don't know but it's a tremendous secret, and I oughtn't to talk of it!"

Hope wished to talk of it, and now she had to unmask. "Was it the tablet he wants to put into the library to his father?"

"You do know about it, then!" Mrs. Enderby rejoiced. "What do you think about it?"

"Why, nobody could have any objections, could they? If his father gave the library building to the town?"

"No, certainly. I fancy they'll be only too glad to have him do it. At any rate, he's going on with it. He's got a sculptor to design it, and as soon as it is finished he is going to have it dedicated here. He hasn't fixed on just the time. Dr. Enderby had a letter from him this morning, saying he had thought of Decoration Day, but that he had consulted with some one in whose taste he had special confidence, and this mystical unknown had suggested to him that it would be taking the day from those whom it belonged to for something else; and he wanted Dr. Enderby to suggest another date not much later. Dr. Enderby proposed his father's birthday, and very likely he will decide on that unless his unknown adviser counsels differently. Do you suppose it is that Mr. Falk who was here with him?"

"I think he would be likely to *ask* Mr. Falk," Hope demurely conceded, with eyes that could not help falling under Mrs. Enderby's.

"Well, whoever it is, Dr. Enderby admires his sense and his feeling." And, at this, the question in Hope's mind whether she should tell Susie Johns about the affair went out of it. She could not do so now without seeming to brag. She was not going to brag, but she felt proud of having the sense and the feeling that Dr. Enderby had praised. "Dr. Enderby liked Mr. Langbrith's frankness, too, in confessing his own want of thoughtfulness."

"Yes, that was nice," Hope said, with some tacit

misgiving for the sarcastic tenor of the letter in her pocket. She said to herself that it was the only way to get along with James Langbrith. If you did not laugh at him a little, he would be unbearable. But she thought that, if she found a letter from him in the post-office, she might not mail hers, at least till she read his.

"Dr. Enderby," the rector's wife pursued, "thinks very highly of Mr. Langbrith. Of course, every one has their faults, but he thinks Mr. Langbrith really tries to overcome his when he sees them, and he bears being shown his weaknesses very well. Dr. Enderby says that is the most uncommon kind of virtue. I didn't quite like Mr. Falk's sarcastic tone with him, but I suppose Mr. Langbrith knows how to take care of himself. Sometimes young men seem to enjoy that. It's like their 'scrapping,' as they call it. But Dr. Enderby says that Mr. Langbrith was just as nice with the cold way Dr. Anther took his plan for the tablet."

"Didn't Dr. Anther like it?" Hope asked.

"Apparently not. He didn't say why, and that made it all the more awkward for Mr. Langbrith. Dr. Anther didn't seem to take any interest in the project, and yet Mr. Langbrith's father was his old friend."

Hope mused darkly for a moment, then she brightened to a laugh. "Well, it doesn't seem to have discouraged Mr. Langbrith very much."

"No, it hasn't," Mrs. Enderby recognized with a laugh of her own, "and I'm glad of it. I think it's a very good plan, and it will be an attractive addi-

tion to the front of the library—so very plain. I
believe in commemorating such things. It helps
to make a place historical, and we have so little
history. But Mr. Langbrith is so very sensitive,
and I don't like to have him hurt. I know he suf-
fers very much when he has found himself in the
wrong."

"Nobody enjoys that," Hope suggested.

"No, of course not; but his ideal is so very high.
He does always want to do what is fine and noble.
I can see that. I think he is *rare*. I almost trembled when you got into that little dispute with him
that night: he's not as quick as you, Hope." Mrs.
Enderby questioned with eager eyes the young face
which masked itself against her pursuit in a smile.

"Oh, it wasn't very serious."

"Not for you, of course, but it was for him. He
was making a brave show, but I could see how very
—very— He isn't as satirical as you are. You
must be careful of that keen little tongue of yours.
Oh, dear, what am I saying? You do forgive me?
But girls don't know how the things they say rankle
in young men's minds, and how eager young men
are to have the approval of girls they respect.
There! There comes Dr. Anther *now*. I wish I
had the courage to ask him why he doesn't approve
of the tablet. Good-bye, dear; I'm going into this
store. Are you going to the post-office? I believe
I'll go with you—or no! If I waited to meet Dr.
Anther, I should be sure to ask him, and I've no
right to. Well!"

Mrs. Enderby slipped into the door-way where she

had scarcely halted the girl, and Hope tilted on towards the post-office with not so light nor so swift a gait as before. It was silly, of course, to mind what Mrs. Enderby said; but she had now fully agreed with herself that she would not mail her sarcastic note to Langbrith till she had seen whether there was a letter. She flushed when the girl clerk gave her a letter from him, and she turned the corner at the post-office to be able to read it unmolested in the by-street leading to Susie Johns'. It was so full of what seemed to her a swelling self-satisfaction that she did not look up the date to see whether it had been written before her last reached him, but pushed it into her pocket, and, hurrying round the square, without stopping to see Susie Johns, she reached the post-office again, and shot the note she had with her into the slot in the door, and walked vigorously homeward, with the full approval of her judgment and a just indignation for her momentary betrayal into a mistaken mercy for an offender so hardened as James Langbrith. She had to pass his mother's house on the way, and she saw Mrs. Langbrith out in the sun before it, stooping to look at the perennials in their bed beside the door. But Mrs. Langbrith did not see her, and Hope got home in a defiance of Mrs. Enderby that kept itself from being articulate with difficulty.

XVIII

Mrs. Langbrith came out every fine day to look
over her flowers at first, and then to work over them.
She made the man clean up round the tall syringas
planted at intervals along the brick walk to the
gate, and about the lilacs that overhung the fence.
She followed him as he combed down the limp last
year's grass and raked the dead leaves and stems
into heaps at the points she chose and then set fire
to them. At tea, she liked to have the dining-room
windows a little open, that the homely smell of their
burning heaps might come in with the fresh evening
air and possess her with the dreams of that girlhood
which now no longer seemed so far past. She
thought Dr. Anther might stop some evening in
going by; but if she caught sight of him in the distance, she went in-doors. She realized that their
embrace at their last meeting was more like a final
parting than a pledge of union, unless she were ready
to do what she wished but was afraid to do. Yet
this thought of it had the greater sweetness for that
reason; and the love that had come into her life so
late was the more precious because it seemed to
have come too late.

Towards her son, grown a man, she felt its inde-
corum in a kind which she could not quite formu-

late, but which was distinct enough. If her love had come when she was younger, and he still a child, it would have been different; and yet she could not blame her friend for not knowing himself sooner. That blame would have been as indecorous towards Anther as now the thought of him was towards her son. Before her marriage her fancy had scarcely been stirred. She had gone the round of the simple children's amusements in her country neighborhood —the parties and picnics and school festivals; but no little boy had been her beau. She had not even been teased by her mates about any one. She was younger in experience than any girl she knew in the mill when Langbrith cast his eye her way, and suddenly, somehow, through her necessity and helplessness, made her his wife. She certainly was not aware of anything like love for him, so far as she imagined love; but she was flattered and dazzled and overcome, and she supposed that she was marrying as other people married, and for the reasons that they had. Her awakening from her illusion was like the terror of a child which has not enough knowledge of the world to match its experiences with those of others. In a fashion not definite or articulate, she accepted her lot as a common lot in wifehood; and, as she had supposed herself to have married from the usual motives, so she now supposed that what she underwent was not unusual. From her sufferings, she formed a notion of marriage grotesquely false, which was like a child's misconception of life, and the spell of this kept her submissive. She did not talk of what she underwent; no one talked to her of such things, and apparently it was not the custom.

Her childlikeness so prolonged itself, not ignorantly, but innocently, through her wifehood and motherhood and widowhood, that, when at last she was aware of liking the man who later loved her, and of trusting him and longing for his affection, it was with a sense of shame as from unprecedented guilt. Before the thought of her son she was so ashamed that she knew she should never be able to tell him of Dr. Anther, nor even allow Anther to speak for himself. She did not feel that her tenderness for her friend could be wrong when she was with him. She was now glad of that sole embrace which they had ever suffered their love, and proud of it; but the knowledge of it sunk her at her son's feet when she imagined his knowing it. Her face burned, and it did not avail her to remember the examples of mothers that had married again, and had lived on with their husbands, and their children by their dead husbands, in unimpaired harmony and mutual respect. She was moved late in her inextinguishable girlhood to her first passion, but only to find herself inexorably consecrated to her widowhood through her reverence for her son's ideal of his father.

At sight of Hope Hawberk tilting lightly down the sidewalk, she was seized with the same impulse to flight as at the approach of the doctor in his vagarious buggy; and she had to conquer far more shyness when, one warm afternoon, Hope caught her so preoccupied with the hired man that it was too late for her to think of eluding her. She shrank together beyond a well-budded lilac, where Hope's gay voice, as if it had a bright, entangling noose of sound, reached her in the chanted salutation, "How do you do, Mrs. Langbrith?" and held her fast. She came reluctantly from her shelter, and advanced slowly towards the gate, on the top of which the girl had laid her arms, and her red cheek for a moment in the hollow of one of them. "Isn't it awfully warm?"

"Yes, it is. Though I haven't noticed it so much, working about. Won't you come in, Hope?"

"Why, I will, Mrs. Langbrith, if you'll let me. I was just coming in, when I saw you." She pushed the gate open and joined Mrs. Langbrith, who turned with her and walked towards the house. "How fast your things are coming on! It seems as if they were twice as forward as ours, and there are twice as many of them. I don't suppose they help each other, do they?"

"I don't believe they do," Mrs. Langbrith answered so literally that it might have passed as a piece of the same whimsicality. "How is your grandmother?"

"She's as energetic as ever. I don't see how she can be. This weather takes all the good resolutions out of me, Mrs. Langbrith, and I don't know how I've got together enough to come and see you. I want to tell you something that I don't want to tell you."

The girl's humor was catching, and the woman caught it. "Well, what is it?" she asked, but she apparently did not expect Hope to answer till she

162

had got her seated at an open window of the parlor, with a palm-leaf fan in her hand.

"Why, it's just this, Mrs. Langbrith. I've got into a scrape with James, and if you can't tell me how to get out of it, I don't know who can."

Mrs. Langbrith's heart fluttered with a varied anticipation, but she united her emotions in the single inexpressive phrase, "I don't believe it's anything serious."

"Yes, it is, Mrs. Langbrith. It's very serious, and it has gone so far now that something has got to be done about it, and I can't have the responsibility left to me."

Mrs. Langbrith listened with the wish for one thing and the will for another, but her will prevailed over her wish, and she kept herself from saying anything leading. She believed that there was some sort of love-quarrel which Hope had come to own, but she was not going to tempt her to the confession. She said, non-committally, "I will try not to hold you responsible."

Hope laughed rather distractedly. "I guess you will have to. It's about that tablet he wants to put up in the front of the library."

Mrs. Langbrith stiffened in her chair, and said, "Oh!"

"Well, James has been writing to me about it since he went back to Cambridge, and I guess he thinks I have been making fun of him, when I was only making fun of the notion that he should take something I said so seriously. Don't you understand?"

"James is apt to take things seriously," his mother said.

"And I'm *not*," Hope retorted, with a touch of resentment, as if she felt a touch of reproach in Mrs. Langbrith's tone, though the words themselves were so neutral. "And that's just the difference, and always will be." The last clause of the sentence was a generality, which the girl seemed to address to herself rather than Mrs. Langbrith. "Now, I'll tell you what it is. He asked me what I thought about his having the dedication on Decoration Day, and I told him I didn't think it was quite fair to take that day from the old soldiers and their families; and he saw it in the same light, and he telegraphed to say that I was right and he wouldn't. And I wrote back making fun of his telegraphing, as if it couldn't wait for a letter."

"I don't see any harm in that. James is very intense in his feelings, but he would see that you didn't mean anything unfriendly—anything—"

"No, of course not. But now comes what I am really ashamed of. My making fun seems to have made him very mad, so mad that he says he is going to give up the whole idea, and won't have anything done about it. He says I have made it seem ridiculous to him."

Mrs. Langbrith cast down her eyes. "James is very sensitive in regard to—Mr. Langbrith."

"Yes, I know that, and that's what makes me sorry. Of course, I didn't mean to hurt his feelings for his father. And now, Mrs. Langbrith, and now—I've got something else to tell you. You know how girls are?"

"Thoughtless, you mean?"

"No — *bad!* Downright wicked! I told Susie Johns about James's telegraphing. I don't see why I should do such a thing. But we were laughing about a lot of things, and that came out. It was as mean as it could be. And now I would do anything in the world to make it right, but I don't suppose I ever can. I don't care a bit about his being mad at me for it; he has a perfect right to be; but what I hate is people laughing at him. I've been to tell Susie not to tell, since I got his last letter, but I know she will. He mustn't give up the idea, because they will say that I laughed at it, and that was the reason, and I am not going to have them. Don't you see? I expect *you* to blame me, Mrs. Langbrith, and never speak to me again; but I shall not care for that if you can think of some way to stop him — to make him not give it up. Why, he *must* go on with it, now. Everybody knows that he was going to do it, and he must. Was there ever such a scrape?"

Mrs. Langbrith sat silent, but this was quite what Hope seemed to expect, and the face that she turned upon the girl was by no means severe, but rather somewhat distressed and puzzled.

Hope went on. "I don't believe it will do any good for me to write to him and tell him he must." Mrs. Langbrith made no comment on this suggestion, and Hope owned, "Well, I *have* written to him, and he's written back, and said that he knows my real feeling now, and he cannot go on. I don't see why he minds my feeling, anyway, and that's the reason why I've come to you. I don't know what made me come to you about it, but I wanted to ask you if you thought it would do for me to ask Dr. Anther to write to James?"

"Dr. Anther?"

"Yes, and tell him not to mind a person who is not worth minding, but to go on and put up the tablet. Tell him that everybody approves of it, and expects it."

Mrs. Langbrith emerged from her absence, but the stare which she bent upon the girl was as silent as her far-off look.

"Will it do for me to ask the doctor? I don't want to do it, because— But I will, rather than let it go as it is. I will do anything. What do you think, Mrs. Langbrith?"

Mrs. Langbrith shook her head, and said, with something that she kept from being a shudder, "Oh no, it won't do to speak to Dr. Anther."

"For me? Or for any one?"

"For you. I—I will speak to him."

"You? Oh, thank you, Mrs. Langbrith! I thought—I hoped—I didn't dare to hope—" The pent emotions, kept in so bravely, broke in tears, and Hope caught her handkerchief from her belt and sobbed into it. "Oh, dear, I don't see why you do it! I don't see how you can bear to look at me, or speak to me, much less do anything I ask you to, after the mischief I've made. But I do, *do* thank you—"

She wavered towards the other, with what design she did not know; but, whatever it was, Mrs. Langbrith put her arms round her, and pulled her head down on her shoulder, and the girl had her cry out there. "Oh, I'm so ashamed, I'm so ashamed!" she kept saying. "I don't know why you let me, Mrs. Langbrith!"

Mrs. Langbrith did not say, and perhaps could not; but when Hope's passion of weeping was spent and she drew away to wipe her eyes, and compose her face, the woman said, irrelevantly, "How is your father, Hope?"

"Oh, much better. I believe the doctor thinks he can cure him."

"That's good," Mrs. Langbrith said as irrelevantly as before, and now she let the girl, with a fling of her arms round her neck, run out of the house unhindered.

Half-way to the gate she met Mrs. Enderby coming up to make a call on Mrs. Langbrith; and, from behind the veil she had caught down over her face, she was able to chant a gay little "Good-afternoon, Mrs. Enderby!" without exciting any question in the lady, except as to how a girl whose life was so tragically conditioned could keep that blithe note in her voice.

XIX

Almost the first thing Mrs. Enderby said was, "That poor, pretty creature, how wonderfully she keeps up!" for this was what was still first in her mind when she took the place at the window which Hope had just left, and looked to see if she could still see her.

Mrs. Langbrith said, "Won't you have a fan?" and Mrs. Enderby thanked her and took from her the fan which Hope had dropped on the table. "It *is* unseasonably warm."

"We often have a hot day like this towards the beginning of May."

"Oh yes, that is true. But the leaves not being out makes it so melting in the sun. Is there the least hope for the child's father?"

"Dr. Anther has always believed his habit could be cured."

"Oh yes, Dr. Anther: how we all depend upon him! Saxmills would be another place without him. We turn to him in so many things. I was just thinking about him—just speaking about him with Dr. Enderby. But, Mrs. Langbrith, what is this I hear about your son's giving up the notion of the tablet to his father? I hope it isn't true—just town gossip."

"James hasn't said anything to me about giving it up," Mrs. Langbrith answered, and she quelled the outward signs of her wonder whether Hope had come to her with a half-confession, and had been twice as silly and light-tongued as she had owned. "He has given up having the dedication on Decoration Day."

"Oh, well, perhaps that's it, and it has got twisted into the other thing. May I say that you have heard nothing from him in regard to it?"

Mrs. Langbrith could truthfully assent to this, but she assented with so much coldness that Mrs. Enderby was struck by it, and a little hurt. In her kind heart, which was equal to most emergencies where excuses were needed for offences, she accounted for the coldness as the expression of rustic shyness. She had known village modesty to take the form of village pride, and, later, unmask itself in touching gratitude.

"We all," she went on, after thanking Mrs. Langbrith for her assent, "think it such an admirable idea, and Dr. Enderby particularly favors it. He feels it so important to recognize *character*, especially when it has influenced a whole community as Mr. Langbrith's has influenced Saxmills, and stamped his traits on the place, as Dr. Enderby says, that I believe he would have been glad to have a tablet to Mr. Langbrith's memory in the church."

At this point Mrs. Enderby certainly expected some sort of response; but Mrs. Langbrith preserved a silence of unbroken iciness. Perhaps she did not

169

like the notion of a tablet in the church. Mrs. Enderby went on:

"But, of course, he feels that there is a peculiar fitness in its being in the library building. We all do, and I am sure every one will be glad to hear that there is nothing in that report, or nothing but a perversion of the Decoration Day part of it."

Mrs. Langbrith made no sign of gratification in Mrs. Enderby's conclusion, and Mrs. Enderby had to go away in an uncomfortable misgiving for the effect of the interest she had shown in the matter. She had no misgiving for the interest itself. That was simply a duty towards one of her husband's parishioners, such as she had promised herself to fulfil towards all after she had so reluctantly consented to his taking the parish of St. Cuthbert's at Saxmills. She felt that she was not only following him into the wilderness—anywhere over twenty miles from Boston was the wilderness for a Bostonian of her elect origin—but she had fears of the peculiar difficulties which a priest of Dr. Enderby's socialistic —she called them "sociological"—tendencies would have in a cure of proletariat souls, housed in a temple built with money from their exploitation. Langbrith had given St. Cuthbert's small but sufficient church to the parish, as well as the library to the town; and Mrs. Enderby's question was whether her husband could keep that perfect conscience between a due sense of gratitude towards the giver's memory and his duty towards the employés of his son's milling property in the event of those differences which might any time arise between capital and labor. She had been wakened by this question one memorable night, and had not been able to wait till morning before submitting it to Dr. Enderby, in a conscience inherited from Calvinistic forefathers through a Unitarian father who had preserved nothing from his ancestral faith but the conscience transmitted to his family of daughters. Dr. Enderby's own conscience was of the same lineage, and it cost them both a night's sleep to decide the point. In fact, it was not until late into the next afternoon that they had reasoned to the conclusion that to do right was his sole duty, and that to shrink from conditions which might sometimes render the right embarrassing or difficult would be a confession of unworthiness for the office they both wished to magnify.

Mrs. Enderby came away from Boston with all the reluctance that she had foreseen; but, though she was followed by the sympathies of her friends, she had as yet experienced nothing which turned her mind towards them in longing for their pity. Saxmills had not proved quite the social desert, beset with dangers, which she had sometimes foreboded. She had there, as everywhere, her husband, first and foremost; and, besides, there were several people she liked. Not counting those she loved because they were poor and sick and dependent, there were, among those she liked, Judge Garley and his wife, who were agreeable mid-Massachusetts town-folk, reasonably cultivated and passably acquainted with life, by reason of several winters' official residence in Boston; and she liked Mrs. Langbrith, ordinarily, very much, though she was not quite what Mrs. Enderby would have quite called cultivated, and certainly not acquainted with life. But her shy charm was a great charm for Mrs. Enderby, and it was much in her favor that she always made Mrs. Enderby think of that old-fashioned, late-summer flower, mourning-bride. The abiding girlishness of the long-widowed, middle-aging woman responded to a girlishness of her own, from which she was fond of all the nice young girls of the village, like Hope Hawberk and Susie Johns and Jessamy Colebridge, and such others as did not dismay her by their fearlessness with the young men. Outside of the mills, the young men were, indeed, so few that there was, perhaps, not much reason to be afraid of them. But, above all, she liked and respected and honored Dr. Anther, whose life had such a daily beauty that she could better have expressed her sense of it if she were still a Unitarian than she could now she was a churchwoman. She was constantly finding him in the houses of affliction, which she visited in her own quality of good angel, and it was without surprise or any feeling of coincidence that she now met him coming to the gate of a common patient, which she opened next after closing Mrs. Langbrith's.

She merely said, "Oh, how delightful, Dr. Anther! I was just thinking of you." And then she added, "I hope you leave our poor sufferer better?"

"*You* will, after you have seen her," the doctor said, shifting his little bag of medicines from his right to his left hand, so as to take the hand of Mrs. Enderby put out to him. He had a fine perception of her lady-world in Mrs. Enderby, and liked to say as nice things as he could to her. "She needs cheering up, and you'll be better for her than my medicine."

"If I could believe you were serious in your civilities, I should be conceited; but I know you only say such things to cheer *me* up—not that I need it just now. I've been to see Mrs. Langbrith, and she has reassured me in regard to a strange report I had heard. I wonder if you had?"

"Better try me," Anther said, twitching his bag up and down with a latent impatience.

"Why, merely that her son had given up the notion of the memorial tablet for the library front. Had you?"

"No," Anther replied, shortly, jerking his bag with open violence.

"Well, if you do, there's nothing in it, as far as his mother has heard. He *has* changed his mind about having the dedication on Decoration Day, and the report probably arose from that." The doctor said nothing, and once more Mrs. Enderby was bruised and disappointed by the bluntness of village manners—this time from one who had always been so responsive. But she rose above it, as she would have said, so far as to excuse him in consoling herself. "But I see your mind is on your next patient, doctor. It's cruel of me to keep you, and I won't any longer. Good-bye!"

She went in to cheer up the sick woman, but even after the exhilarating effort she came away with a little lingering impression of Dr. Anther's indifference to her news. She submitted her impression to her husband, whom she found struggling with a sermon of an hour's length rebellious to his ideal of twenty-five minutes. He detached himself to examine the impression, and to match it with one he had brought away from the supper at Mrs. Langbrith's, when Dr. Anther had received young Langbrith's proposal of the tablet with so little interest. "Perhaps," he suggested, "it is not that country uncouthness altogether; there may not have been all that friendliness between the doctor and the elder Langbrith which we have inferred from his present relations to the family."

"Why, have you heard anything of that kind?" Mrs. Enderby was of an eagerness in her inquiry which her husband thought it well to repress.

"No, nothing at all. It's pure conjecture with me."

"It would be very interesting." Mrs. Enderby sighed for the evident want of foundation in fact.

"Yes, but, Alice, don't let it take possession of your fancy. It would be very unjust and it might be injurious."

"Oh, I should not dream of mentioning it to any one. To whom could I? But it happens to tally— is that slang?"

"I don't know that it is."

"I oughtn't to use it if it is, in a place like this. It happens to tally with something that has come into my mind. I have always wondered why Dr. Anther doesn't marry Mrs. Langbrith."

"He may not wish it, or she may not."

174

"It would be such an appropriate thing. They are old friends, and they are not too old. Her son will soon be leaving her—I know he's in love with that poor, pretty, joyous Hope Hawberk; and the doctor must have always been very uncomfortable at Mrs. Burwell's, and now she's going to break up, and where *will* he go?"

"He certainly might do worse than go to Mrs. Langbrith's," the rector allowed; "but still there is no more proof that he wishes to marry Langbrith's widow than that he bears a grudge to his memory."

"No, but don't you see that, if he *did* want to marry Mrs. Langbrith, it would make him willing to have Mr. Langbrith forgotten? Wouldn't that be natural?"

"I'm afraid it would, Alice," the rector said, with a regretful recognition of a trait of fallen man.

"And wouldn't it account both for the way Dr. Anther behaved that night and for the way he behaved just now?"

"It might; but don't you see we are proceeding upon a pure hypothesis?"

"That is true," she consented; and now, having not before removed her hat, she pulled out the long pins that pierced its sides into the mass of her handsome graying hair, and lifted it off, carried it out of the study on her hand, thoughtfully considering it as she went.

"Of course, Alice," he called after her, "we must both be careful to keep our hypothesis to ourselves."

"Oh yes, indeed! I shall be very careful not to speak of it."

175

XX

After indulging his resentment of Hope's ridicule to the violent extreme of renouncing all intention of the memorial tablet, Langbrith allowed a natural revulsion of feeling to carry him so far back as a renunciation of Hope instead. He wrote her an angry letter, in answer to hers asking him not to mind anything she had said for the reason that she was not worth minding herself. Then he felt so much stronger that he returned to his intention, and got Falk to go with him into Boston to the studio of the young sculptor who was modelling the bas-relief. It had to be done from few and rather poor photographs, for it had not been one of his father's excesses to sit often for his picture. There were some ferrotypes and still older daguerreotypes from which the sculptor had imagined a head more or less ideal. Falk tacitly considered the ideal an improvement on the portrait in the library at Saxmills, and he had kept Langbrith from sending for the painting by sufficiently offensive censures of its woodenness. Besides, as that was from a photograph, too, he held that there would be no advantage in studying the tablet from it.

The young sculptor was a find of Langbrith's in the course of his own æsthetic development. He had seen some idealistic banalities of the artist in an art-dealer's window, and had liked them so much that he had got Falk to come and look at them, too, and then join him in looking up the sculptor, who, when looked up, proved to be a beautiful, poverty-stricken young Jew, with black hair bushing out over a fine forehead, and, under the forehead, mobile, attentive eyes. He had a profile more Hellenic than Hebraic, and cheeks and chin already blue from shaving a dense beard. It appeared that he had made the banalities to sell, and that he could do stronger if not truer things, as the casts in his studio witnessed. He entered into the motive of the medallion, as Langbrith presented it, with an ardor that matched Langbrith's, and he roughed it out in the clay so quickly that in twenty-four hours he had something to show his patron. He had conceived so aptly of his patron, if not of his subject, that he flattered the effigy of the elder Langbrith into a likeness of the son, who stood before it in content little short of ecstasy.

"Falk," he said, "it has the ancestral look—the look of race. I can see myself in it. That must have been the way my father looked. Wonderful!"

"It *is* like you," Falk said, with a glance at the sculptor, who was watching Langbrith with subtle and shifty eyes. "I always supposed you rather resembled your mother."

"Not at all," Langbrith retorted. "There may be something in our features, but her expression is totally different. I see nothing of my mother in this."

"Well, if you're satisfied, that's the end of the story."

"But look, Falk, look at these old pictures!" At the first question the sculptor had supplied them, and Langbrith now held them in his hand, studying them and then the sketch. "You can see that the outline is the same, and Mr. Lily has read the character into the face which these caricatures belie. It's the artistic resurrection from the mechanical death of these tintypes. It's miraculous!"

"Well, Mr. Lily probably believes in miracles."

The sculptor presented an impervious surface to the smile of Falk's irony, and Langbrith continued to effervesce without heeding his friend.

"And you've taken my notion most delightfully about the inscription, Mr. Lily. It's just the effect I wanted, fronting the eyes there in those lines of compact capitals, and balanced by that low relief of the mills at the back of the head. Of course, I know that there's nothing definite in those details yet; but the face is so perfect, so struck as if with a die, that I dread to have you touch it. Do you think you can keep just that look in working it up?"

The young sculptor pouted his handsome, thick, red lips, and said, "I think so," stealing a glance from his subject to his patron.

"Well," Langbrith sighed, "I shall have to trust you," and Falk laughed out. "What's the matter?" Langbrith demanded.

"Of course you'll have to trust him! You're not running this work of art."

"Oh, I didn't understand you. Of course! And how soon can you have it done, Mr. Lily?"

"How long can you give me?" the sculptor asked.

"I did think of Decoration Day for the dedication, but I've changed my mind about that. I think now I will have it on my father's birthday—the 29th of June. Could you have it ready by that time?"

The sculptor seemed considering seriously, and at last he said, reluctantly, as certain people do to enhance the value of a concession, "I think I can, if there's no delay in the foundry. If there is, you know you can dedicate a gilded plaster copy and put up the bronze later—any time."

"Ah, I don't believe I should like that. It wouldn't be in character with my father. He always paid cash. I shall trust you to have it ready in time. I know you can have it done."

Langbrith remained studying the sketch until Falk's restiveness obliged him to break from it. The more he saw himself in his portrait of his father, the better he was pleased, and the truer he decided the likeness to be. The family look certainly was there, and what greater truth could he ask? With all the self-satisfaction of the academic side of his nature he rejoiced in what he decided to be an ideal presentation of his father's face.

The sculptor followed him and Falk to the door of his studio, and bowed them out.

"Little Sheeny!" Falk observed, when the door had closed upon them.

"He says he is an Italian, but he's all the finer

artist for his drop of the 'indelible blood,'" Langbrith said, still rejoicing.

"Every drop of blood in his body is indelible. He ought to be a puller-in. He could sell *you* any misfit in the store. I'll do a puller-in at a Roman statuary's for *Caricature*, and I'll have this fellow working off a Mercury on you, come up from your Sabine papyrus-mill."

"If you didn't like it, Falk—"

"Why not say so? What good would that have done, with your infatuation? Besides, I didn't say I didn't like it. There's a lot of infernal *chic* about it. In its way it's damnably good, but it's you, you poor innocent, right over again, and that's what he was aiming at from the first moment he got those eyes of his afloat on you."

"If that's the way you feel—" Langbrith began, half turning.

Falk caught him by the coat lapel and pulled him round again. "What are you going to do? Countermand it? You couldn't get a better job, and the poor devil needs the job. Don't I tell you it's good? It's all the better for having so much of you in it, if I do say it that hate to. Come! you've blundered and he's swindled, into the very thing. Let him go on!"

Langbrith moved reluctantly forward. "If I could trust you, Falk—"

"You've got to."

"I never can make you understand how I feel about my father. He's a religion with me, and anything that seems to belittle him or belie him is a profanation. If I didn't feel that somehow the fellow *had* got my father into the thing, I wouldn't let him go on. Of course, I can see that he has had to work back from me—get the life into it from me! But, apart from the question of the likeness, don't you think it's good?"

"Haven't I said so? The fellow has done it mighty well."

"He has taken all my suggestions," Langbrith said, reaching out for a little more kindness.

"Yes, and Saint-Gaudensized them. That's all right. That was the thing to do. At this time of day he couldn't get away from Saint-Gaudens."

"I'm anxious to have him go on," Langbrith dreamily continued, "because he can get it done in time, and I don't know who else could." He hesitated, as he must, even in the intimacy of his confidence with Falk, before adding: "I've just told Hope that I'm having him do it. I've told her that I had taken it up again."

"Why, had you dropped it?"

"Yes," Langbrith owned, uneasily. "We had a little misunderstanding about having the dedication on Decoration Day. Falk, I want to tell you; but you're so sharp—"

"Oh, go on. I'll spare you on condition that you're honest."

"Well, you know she didn't approve of that idea, and she put it in such a light about its not being fair to take the day from the old soldiers whom it belonged to that I saw it just as she did, and I telegraphed her—"

"Telegraphed!" Falk opened his mouth for a laugh, but shut it again without laughing.

"Oh, laugh! I don't care now! And she came back at me with a letter that cut me to the quick. She's terribly sarcastic, or can be. She made all sorts of fun of me for thinking my agreement with her opinion so important that it could not wait for a letter. Of course, she put it in the way of mocking at herself, saying that she never supposed before that she was of so much consequence that her opinions had to be accepted by telegraph. It ground me awfully, and I took it like a perfect ass."

"Naturally!" Falk interjected.

"Oh, don't mind *me!*" Langbrith exulted. "I wish I could always be an ass to such purpose. I wrote back to her that she needn't be anxious hereafter, for I had given up the whole scheme of the tablet, and I should never trouble her again by my method of asking or acknowledging her opinions."

"That sounds so wise that it must be true," said Falk. "Go on. I thought I knew you, my young friend, but you are unfathomable."

"I begin to believe it. And that brought a letter from her, protesting against my giving up—very dignified and impersonal, and all that; but I was still so sore that I let her letter go unanswered a couple of days, and then there came another from her, entreating me to go on with my plan, and saying if I didn't she never could forgive herself; that she had not meant anything by what she said of my telegraphing; that it was only fun, and I oughtn't to take it in earnest; but if I must, she withdrew it all. She said she knew that if it got about that I was going to give up the plan it would make all sorts of talk and be very disagreeable."

"She had probably told about your telegraphing to some of the other girls," Falk interpreted. "Susie Johns, likely. Or even Jessamy Colebridge."

"That was what I thought at first, and it made me madder still."

"Yes, you are that kind of ass," Falk assented.

"So I decided not to answer that second letter. And then came one that was fairly imploring. It was dated a day later, and she said in it that she was so miserable she had to write again, and she should not rest till she heard I was going on with my plan."

"You must have felt proud."

"No, I didn't. I felt ashamed."

"I'm surprised."

"Oh, I deserve anything you can say. She kept up the effect of joking, but she must have been serious, or she wouldn't have written three times."

"And having brought the suppliant to her knees, what did the Prince of Saxmills graciously deign to answer?"

"The P. of S. had already answered the second letter, on second thoughts. He had written to tell her not to think of the matter again; for, on looking it all over, he found that he couldn't relinquish the plan now. He tried to make his decision seem unrelated to her, because he didn't think it fair to take the advantage she had given him."

"I'm glad to hear it," Falk said. "It's a little fact that enables me to continue your acquaintance, which I was just going to drop. Langbrith, don't you know that that girl is one of the most—"

"I think I don't need any one to tell me what she is," Langbrith interrupted, haughtily. Then he instantly stooped from his height. "But I appreciate your feeling, and I thank you. I was glad that I had answered her second letter, and that my answer and her third had crossed, because I got a letter from my mother with the third from Hope, saying Hope had been to see her, and had told her the trouble she was in. I shouldn't have liked her to think I had done for my mother, even, what I wouldn't do for her."

"Yes, you saved your distance. I congratulate you."

"That's mighty kind of you," said Langbrith, fondly. "And you think—you think, don't you, that the whole situation looks rather favorable for me?"

"Oh, come now! I can't go into that, Langbrith. It's more than I bargained for."

Langbrith went on dreamily: "She must have been a good deal troubled. She asked my mother how it would do to get Dr. Anther to write to me, and my mother had to put her off by promising to ask him herself. Afterwards she decided not to ask, but to write to me instead. I rather wish she *had* asked the doctor," he concluded, meditatively.

"Why?"

"Oh, there is something that has rankled in me

184

ever since the night of our supper-dance—the way Anther took my proposal of the tablet. Didn't you notice anything peculiar in his manner?"

"He didn't seem to take a great deal of interest," Falk owned.

"He is the oldest living friend of my father. I didn't like it. I had it out with my mother, the next morning, and perhaps that made her reluctant to ask him to take any part in it. She is very proud where there is any question of slight to me, and I suppose she felt as I did about my father. Perhaps she found she couldn't do what she promised Hope. But that's a small matter. The great thing is that I hadn't waited for her letter before writing to Hope. I can't be too glad of that. I should have hated even to seem to have done for another what I hadn't done for her. I'm hard hit, and you know it, Falk, and now you've got to listen. That girl is the rarest human creature on this earth."

"Why not say in the universe, and be done with it?"

"Because I don't want to wrong her by any sort of extravagance. She's so perfect, she has such poise, such spiritual proportion—through her sense of humor, I suppose—that any sort of excess seems an insult to her. She's wonderful: I see that more and more. Sometimes, when I think of her hard life with that opium-eating father of hers and that belated old Puritan of a grandmother, and how her days must pass between the horrors of his narcotic and her religious frenzies, I wonder she can keep her sanity. But she is the sanest and sweetest and wholesomest and loveliest soul alive. She doesn't seem any more related to her surroundings or origin than the singing that you hear come out of a church window in summer, and that you can't connect with the stifling and perspiring congregation inside. It's disembodied worship, and she's just joyous girlhood incarnate.''

"Oh, Lord! I can't stand much more of this!" Falk groaned.

"When I think of her," Langbrith went on, careless of his sufferings, "I seem to myself the most contemptible and unworthy caricature of humanity —full of every kind of ridiculous imperfection and detestable defect; a wrong-headed, stubborn mule, with an instinct for kicking at the wrong time and in the wrong place, with a hide so thick and a fibre so coarse that any suggestion for its own good, short of a big stick, is lost on it."

"Well, now," Falk got in his revenge, "you can understand just how you seem to other people a good deal of the time when you're *not* thinking of her."

XXI

"WHAT's this I hear?" Hope's grandmother required of the gayety which, even beyond the girl's
cheerful wont, marked her rebound from her trouble
after Langbrith's second letter came. "Folks are
sayin' that James is not goin' to put that inscription
of his father on the library, and then again that he
is. You know anything about it, Hope?"

"Yes, I know all about it, grandma. They're
right both ways. He wasn't and he is, unless he's
changed his mind again."

"How do you know?"

"He's written to tell me."

The old woman's eyes flared on the smile in the
girl's. "What's James Langbrith writin' to you so
much for?"

"He seems to like to."

"You engaged to him?"

"Not at present, grandma."

"Better see 't you ain't. I don't like the breed
any too well. I hain't ever been satisfied at the
way he got your father out of the business, and your
mother wa'n't at the time. But she's in her grave
now, and your father can't ever be got to say a
sensible word about it; just praises him up, if you
try to talk with him, when everybody knows that

Royal Langbrith never drew an unselfish breath. He never went to church and he never darkened anybody's doors in the place, and why he gave that library to the town nobody will ever know. It wa'n't like him to give anything; and I guess if your father could be got to tell the truth once, folks would sing a different song. I don't like your close-mouthed kind, that force one out of partnership and never say why nor wherefore."

A yell from the chamber in the half-story over the room where the two women sat at breakfast offered itself an apt explanation. Groans and sighs followed, and gasps of prayer and thanksgiving, and then muffled fumblings and stumblings on the floor, as of a man getting out of bed and dressing.

"My! how it does always go through me!" the old woman quavered. "I don't see how you can take it so, Hope! I can't seem to get used to it!"

"I was born used to it," the girl answered, with a patience that was cheerful, even smiling. "He hasn't had such a dream for a good while. He must have been at the laudanum bottle, instead of the other. I'll just look." She ran quickly up the crooking stairs, and her voice made itself heard in fond reproach. "Now, father, how *could* you? What's the use of my trying to trust you? And don't you see what it does? Just throws away all the good effects, and brings us back where we were before."

"Yes," she reported triumphantly to her grandmother, as she reappeared with a large, empty bottle in her hand. "It's just as I supposed, and I've got to go and tell Dr. Anther as soon as I've taken father some coffee. Will you put it on the stove, grandma, while I help him dress? I must hurry."

She found Anther at his office, after his somewhat later breakfast, and he could not refuse to smile when she reported the fact in its humorous aspect, with an unbroken trust in the fortunate result.

"I guess we've got to begin again, Dr. Anther," she said, uncovering the empty bottle. "You see what father's been doing!"

"How came you to notice?" he asked.

"Oh, a dream that almost raised the roof," she laughed. "I should have thought it was the skeleton-man and the green dwarf both, but I didn't ask. When I found this, it wasn't necessary." She offered him the bottle, which he received with a face losing its sympathetic cheerfulness. Her eager nerves took alarm at his gravity. "You don't think he's worse?"

"Oh no"—the doctor came back to his professional reassurance—"it's a little disappointment when we had got him so far along, that's all. But it's not a thing to discourage us. We shall have to begin over again, as you say." He set the bottle aside. "I'll bring it to him, and have a talk with him."

"Oh, do!" the girl said, back in her gayety again. "Your talks do him more good than the medicine, I believe. I wanted to have a talk with you myself the other day."

"*You* haven t taken to laudanum, I hope," the doctor said, returning to his smile.

189

"Well, it was something that gave me bad dreams
while it lasted. I thought I had made mischief,
and made it out of pure silliness." The doctor's
smile took on the incredulity that prompted her to
go on. "Yes, I did, I made mischief—set the gossip going. You know you have heard it, Dr. Anther—about James Langbrith giving up putting the
tablet to his father, and then deciding to do it?"
The doctor reluctantly assented. "Well, *I* did
that." She possessed him, laughing and blushing,
of the whole case, and then waited confidently for her
acquittal, or, rather, went confidently on without
it as something that might be taken for granted.
"Before James decided to do it, finally, I was so
worked up that I went to Mrs. Langbrith, and
coaxed her to ask you to write to him and tell him
to go on. She promised, but concluded to write
to him herself. By that time he had made up his
mind to go on, anyway. You see what a narrow
escape you had. I thought I ought to tell you."

The doctor said, "Yes, that's right," and then
a vagueness came into his gaze that made the girl
laugh.

"Well, I'm going now," she said.

"Oh, I didn't mean that."

"Well, I'm going before you *do*."

"Hope, you're a good girl," Anther said. "I
mustn't praise you to your face; but if any one ever
wonders to you that you can keep up as you do,
you tell them I say they don't know anything about
it; that there isn't one in ten thousand that could
bear as you do what you have to bear; but don't

ever get to supposing that it's your duty to be sad about it. It's your duty to be gay."

"Well, that's what I like being, you know, Dr. Anther. It's so easy that it doesn't seem like very much of a *duty*."

She had risen, and she stood prettily smiling at him; and he looked at her, and then suddenly turned his back on her, as if shunning a temptation. He longed to take her in his arms. "Well, I'll be up in the course of the forenoon."

Before he went to see Hawberk he dropped his buggy anchor before the Langbrith mills, and found his way through the works, where the odor of wash-day from the pulp-vats was only denser than it was outside, to the office partitioned off in a corner of the building. He pushed open the door, which closed with a weighted cord, and shut himself in with John Langbrith.

The manager was sitting at his desk, and at the opening and closing of the door, which, through the shuddering and muttering of the machinery, made itself seen rather than heard, he got lankly up and took the doctor's offered hand, which he pushed horizontally back and forth without looking at him. "Good-morning, doctor," he said, and then he glanced at the papers on his desk with a desperate sigh.

"John Langbrith," Anther began, at once, "you know about this scheme of James's for putting up a tablet to his father?"

"I've heard about it. I heard he had dropped it."

"He's going on with it."

"Well, what have I got to do with it? I've got enough on my hands looking after my job here in the mills."

"Has he consulted you about it?"

"No."

"Well, I advise you to write to him and urge him to stop it."

John Langbrith lifted his narrow, yellow eyes and met the doctor's. "Why?"

"You know what your brother really was."

"So do you. Why don't *you* tell James to stop it?"

"That's nonsense. Sometime the truth must come out."

"You mean that you will give it away?"

"That's stuff! But there are others."

"Hawberk? What's *his* word worth?"

"Nothing now. But if he pulls up—"

"He'll never pull up."

"I have hopes of curing him, and I tell you that, when the truth comes out, there will be shame and sorrow for that boy and scandal for the community. It isn't for me to tell him about his father, and you know his mother can't. It's for you, or for Hawberk."

"He's making up to Hawberk's girl, ain't he? Then they can fix it with Hawberk. My job is to look after the mills. I'm not going outside of it."

"It would be an outrage to let that girl marry the son of the man who ruined her father, and you will be a partaker in the wrong unless you speak. Tell James about it, and let him act from his instincts of honor and justice."

"I can't go outside of my job."

"If he's allowed to go on and marry that poor girl, it will be taking a cruel advantage of her. She will be marrying him blindfold. She will be trapped and fettered and manacled for life."

"I can't go outside of my job."

"When the truth is known, and it must be known, the effect with the public will be hardening and depraving beyond that of any bad life openly lived. It will breed a spirit of defiant cynicism, and put a premium on hypocrisy. It will be inconceivably debauching and corrupting. Think it over, Langbrith!"

"I sha'n't go outside of my job."

The words came with an unexcited dryness which convinced Anther of their finality, and kept him from saying more. But Langbrith followed him to the door with words more forbidding still.

"I don't owe that young man anything; let him make a fool of himself any shape he wants to. And I don't owe this community anything; it may rot for all me. And I don't owe *you* anything; you mind your own business! I don't want you bothering round here any more."

Anther made him no answer. He did not blame him greatly. He knew John Langbrith to be as clean a man as his brother had been foul; but he knew that it was not in the measure of his narrow nature to do what he had required of him. His job was the measure of him, and Anther owned to himself that John Langbrith could be safe only in keeping to that.

He was lifting his hitching-weight into the buggy when he felt a hand on his shoulder, and recognized in its touch the heavy mental method of Judge Garley. "Been asking brother John for help?" the judge conjectured.

With the weight dangling by its strap, Anther said, "Yes, I have."

"Well, you didn't get it, I presume. Brother John likes the safe side, which happens in this case to be the inside. I have been considering the matter you laid before me the other day, and my advice is to drop it."

"I sha'n't drop it!" Anther answered, sharply.

The judge did not mind his wilfulness. "At this late day, nothing can be done—nothing but mischief can be done—by drawing the frailties of our departed brother from their dread abode."

"That's what you said."

"Well, that merely proves that I saw it in the right light at first, before taking time to reflect upon it. It increases my respect for myself without diminishing my regard for you, my dear friend. You can accomplish nothing whatever by the course you propose to pursue. An exposure would come from you with a peculiarly bad grace; it is hardly necessary for me to say why; and it would only convince the young man that you were his father's enemy. You could not count upon his mother's corroboration in such an event."

"I could count upon her truth, in any event."

The judge slowly shook his large head. "Not if she is the good woman I take her to be." Anther nervelessly dropped the hitching-weight, and it fell so near the judge's foot that he looked down at it, though he did not move. "My dear friend, you would stand unaided and alone, and the outraged sentiment of the community would be against you. The sentimentality of the community would overwhelm you. Your exposure of the boy's father would be attributed to the worst motive, in the absence of any, and the tide of pity for him would bear you down."

"Look here, Judge Garley," Anther said, "would you deliver the dedication address, if you were asked, knowing what you do?"

"There will be time to consider that point when I have been asked. I may say in general terms that I would refuse to do nothing that I was required by my sense of public duty to do."

"It's a pity you went out of politics," Anther said, dryly. He got a new grip of his anchor-strap, and lifted the weight into his buggy. Then he lifted himself in.

But the judge laid a detaining hand on the frame of the lowered top, and questioned with an anxious smile: "Anther, I hope you are not about to do anything precipitate?"

Anther braced himself for an angry reply, and then fell back against the seat. "Oh, precipitate! I don't know that I'm going to do anything at all."

"Well, that's right," the judge said, and let him drive away.

XXII

THE twenty-ninth of June was fixed for the dedication of the tablet, and the time that passed before that date seemed by no means too long for the work of preparation. The young sculptor, under the inspiration of Langbrith's frequent visits, worked with such ardor that he finished the bas-relief early enough to send the model to Chicopee, and have it cast in the bronze which alone satisfied Langbrith's sense of the sincerity essential in the tribute he was paying to his father's memory; and Falk owned that the sculptor had done his work well. He had done it with a touch that suggested the most modern sculpture, and yet preserved a sort of allegiance to the stern Puritan nature of the subject. Royal Langbrith was there not only in the life, but in what his son felt to be that high personal character proper to him. Here was a man, not of the immediate moment, but of that hour of the later eighteen-sixties which created the immediate moment; the hour of the Republic's supreme consciousness, when all the American forces, redeemed from their employment in the waste of war, were given to enterprises which have since enriched us, and, under the direction of such captains of industry as Langbrith's father, have pressed forward to the commercial conquest of the world. The face, which the sculptor had imagined from the son's face more than from the likenesses supplied him, wore not the old-fashioned Websterian frown of the ante-bellum Americans, when there was no greatness but political greatness in the popular ideal, but had almost an eager smile, full of business promptness, and yet with refined intelligence, a sagacity instantly self-helpful, but ultimately not unkindly. The son's heart glowed within him as he looked at it, and he offered it the ancestor-worship of a man proud of his race, of a dreamer idealizing the future from the past. He wished Hope could see it with him, and the wish reddened him with a conscious blush.

He wrote home to his mother, declaring his entire satisfaction with the work, and predicting her own; and he betrayed his impatience for the event which should appeal with that sculptured face to the gratitude of the community at Saxmills. During his childhood and boyhood, when he had looked out upon the place always as through the windows of his father's house, with a sense of being in it but not of it, he had nourished the arrogant, yet affectionate, longing to dominate it by winning its kindness for himself and his name. His impassioned reveries abounded in dramas of his acceptance by the matter-of-fact little Yankee town, in a sort of seigniorial supremacy, which should be its voluntary acknowledgment of what the Langbriths had done for it; and during the absences of his college years he had not wholly lost this ambition. His temperament had kept him from great knowledge of the world, and such knowledge as he had grown into had given his boyish fancies practical shape rather than destroyed them. He might be a disagreeable fool, as he often approved himself to his acquaintance, but he was not finally an ignoble fool.

At the bottom of Langbrith's heart still rankled the obscure resentment for Dr. Anther's obscure indifference to his scheme that he had instantly felt when he first spoke of the scheme before the village magnates in his mother's house. The bruise of that obstruction against which he had so unexpectedly struck remained, and nothing could assuage the hurt but Anther's conviction of wrong and his confession of it. He wondered at times if his mother had ever spoken of the matter to Anther. He had peremptorily forbidden her to do so, in the first letter he wrote home afterwards, but he had hoped she would. Yet no word came from her concerning it, and he could only suppose that she had too faithfully obeyed him. At times, he questioned his own impressions of the fact, and doubted whether it happened, with the significance which his veneration for his father and his affection for Dr. Anther both gave it; and again he could not rid himself of the belief that it had happened in the form and meaning which it first seemed to have.

Before it happened he had imagined asking Anther, as foremost of the Saxmills men who had known his father, to deliver the dedicatory address; but, with this bruise, this doubt, in his mind, it was impossible to do that; and he felt himself less able to demand the explanation from Anther which he sometimes trembled upon the point of asking than to turn to some one else for the address. He would have preferred Anther to all others, even if Anther had not been his father's old friend; for the doctor had his repute as a speaker of simple effectiveness; his oration at the celebration of the first Decoration Day after the great war was remembered still in Saxmills with the exaggerated admiration which history compels when it becomes tradition. It seemed to Langbrith that no one could do such justice to the quiet, almost disdainful virtues of his father as the quiet, almost disdainful powers of his father's friend. But now he had to devolve for the office of orator upon Judge Garley, a speaker of most respectable gifts, but pompous and ponderous, and of a personal ignorance of the man to be commemorated which, in Langbrith's estimation, all but disqualified him. The sweet in the bitter was the hope that Anther might feel the slight of being passed over, and be duly humiliated; but this did not so much console Langbrith as it might if he had not been hurt in his love as well as his pride.

The judge met the doctor driving through the town the day after he had Langbrith's letter requesting him to make the address, and he overcame a certain embarrassment he had in telling his old friend of it.

"I congratulate you," the doctor said, with ironical dryness; but he did not ask the judge if he had consented.

"I do not know," the judge said, "whether you will approve of my accepting the invitation."

"Oh, approve!" the doctor said, with deprecation which was also ironical.

But the judge showed no resentment. "I didn't think it was fair to bother you with the matter, or else I should have come to speak with you before writing. But I did not see how I could decline, and I believe you will be satisfied with the manner in which I shall treat the subject."

Anther, if he was too much vexed to try penetrating the reserve which the judge's words invited him to explore, felt also that he had no right to take any tone of censure with him. He said, "You couldn't refuse without wounding the boy's feelings."

"That was what I felt," the judge answered, with relief. "I might have pleaded an excuse of some kind, such as intended absence from the place, but I did not like to do so, in view of the fact that I shall be detained here by some business that is coming up at the time. He asked for an early answer, so that he might get ready some biographical material he wishes to supply me with."

There was a twinkle in the judge's legal eye, and a smile at the corner of his legal mouth, and he responded with a laugh to the doctor's remark: "In addition to what I have given you?"

"Yes; I need all that I can get on account of that!" The judge roared at his own fun, and Anther drove slowly away at the jog-trot which was his horse's habitual gait when they were both absorbed in thought. Their heads hung down with the same droop, and the horse looked as if he might be revolving the same distasteful thought as the doctor, with the same sense of helplessness.

Within the week that followed Anther was stopped at different times in his progresses through the main street of Saxmills by different leading citizens, who invited him to consult with them upon points of the common interest. James Langbrith seemed not to have rested, after getting Judge Garley's reply, before addressing himself to the selectmen, the high-school principal, and the Sunday-school superintendent, as well as the chief officers of the Sons of Pythias and the Saxmills Cadets, inviting their co-operation in the ceremony which he had so much at heart. Each of these dignitaries now addressed himself to Dr. Anther, in his succession, with the confident belief that Dr. Anther, as the oldest friend of Royal Langbrith in the community, and as the close friend of his son and widow, would be most concerned in the affair, and would perhaps have some inside authority and information to impart. He had necessarily to disappoint their hopes, but he found himself putting on more and more the air of at least civic sympathy, which they seemed to demand of him. He could not, indeed, show them his real mind without awakening a suspicion he was far from wishing to rouse, without starting gossip that would grow into scandal, and involve the Langbriths and himself in mischievous conjecture. He carried his compliance with their obvious expectation to a point where it became almost intolerably irksome, without seeing the point at which he could refuse compliance. When it came to Mrs. Enderby's calling gayly to him from the sidewalk, and halting him, like the rest, to announce that the rector had just had a letter from young Mr. Langbrith asking him to take part in the dedicatory ceremonies, Anther's soul rose in insurrection. "But you knew he had written, I suppose," the lady said.

"No, I certainly didn't," he answered, with a sharpness which suggested to her the possibility that the doctor resented the young man's not consulting so old and so near a friend, but suggested it not so forcibly as to withhold her from saying:

"Yes, he has asked Dr. Enderby and Father Cody, and Mr. Alway of the 'orthodox' church"— she said "orthodox" with the effect of humoring local usage, but also of putting the word between quotation marks—"all to take part. I believe Father Cody is to ask the blessing, and Mr. Alway is to make the opening prayer. Mr. Langbrith has asked my husband to say something from the altruistic stand-point, as it bears upon what his father did for labor in his time by profit-sharing, and, incidentally, if he pleases, to draw any lessons as to character-building from the example of his personality."

Mrs. Enderby ceased obviously reporting Langbrith's diction, and continued: "Of course, he is rather vague about what he really does want, but Dr. Enderby found his wish, so far as he imagined it, rather suggestive, and he said at once that he would like to talk with you about it. When could he see you, at your entire leisure?" she could not help pushing officiously on, though she had no authority from her husband to ask the question. Anther did not know what to say, between his ire and his embarrassment. In his hesitation she added, "I know how difficult it will be for you to fix a time, but I knew how interested you would be."

"Thank you. I'm afraid I should be very little use," Anther began, but she broke in upon him, to make reparation for James Langbrith's strange thoughtlessness, and to soothe the doctor's wounded pride:

"I'm sure Mr. Langbrith will write you about it. I'm surprised— He probably knows how pressed you are, and wanted to save you all the trouble with details that he could. I rather like his going forward, and doing it all himself, don't you? It shows such spirit, and such a pride in keeping it in his own hands!"

"Yes, yes," Anther said.

"But, of course, you don't have to wait for a direct application from him."

"No."

"Why!" she started in self - surprise. "Why shouldn't you stop in any evening, and have tea with us, and talk the matter over with Dr. Enderby, then? I should so like to hear you two discussing the civic and social significance of such a man as Royal Langbrith, and getting at the psychology of him. You *will* come, won't you? Won't that be the easiest for you? Will you come, say, to-morrow night?"

"Not to-morrow night. I can't fix the time just

now. But I will see. Will you excuse my hurrying off to a patient—"

"Why, of course! How thoughtless of me! But any night will do, doctor, so that it's soon. Goodbye! good-bye!"

She turned from the gate where she had stopped the doctor, and went indoors to her husband.

"I think it's strangely thoughtless of James Langbrith not to have written to Dr. Anther about the measures he's been taking. The doctor feels it, I know, but he's so large-minded that he'll not let it interfere. He's coming here some evening to talk Royal Langbrith's personality over with you."

"Where have you seen him?"

"At the gate, just now. But I didn't call you, because I didn't want to interrupt you. I've told him all about it, and he's coming the first evening he can. I told him any evening would do. I knew you'd want me to. And all I shall ask is to sit by and hear you two analyze Royal Langbrith. With the scientific stand-point which Doctor Anther can supply, and the philosophic and religious view which you can give, I think it will be one of the most intensely interesting things that ever was. Don't you?"

"I hope you'll find it so, my dear. But, really, young Langbrith's oversight seems an extraordinary—"

"Yes, doesn't it! I hardly know how to account for it, but if the doctor can overlook it, *we* can, and he's evidently disposed to overlook it. At any rate, I shall 'keep right round after him,' as the country people say, till he redeems his promise."

She so far redeemed her own promise as to halt Anther, whenever she could reach him, by hailing him with her voice, and, when she could not, by waving him to a stand with her fluttered handker-chief. But it was not till she had almost lost faith in his large-mindedness, and had many times sided with him and against him in his imaginable resent-ment of young Langbrith's neglect; it was not till the eve of the day fixed for the dedication ceremo-nies that Anther appeared at the rectory. He came too late for tea; and, when he did come, he did not invite Mrs. Enderby's presence at the psychological analysis of Royal Langbrith's personality, which he did not enter upon till she no longer had the least excuse for not leaving him to her husband.

XXIII

After Mrs. Enderby went out Dr. Anther remained in a silence which the rector could not quite bring himself to break. He thought that his visitor looked fagged, and that he looked even more sad than fagged. He would have liked to ask Anther about Hawberk, in the way of a beginning, but somehow he did not, though he had heard that Hawberk was holding up a little, and he was interested in the experiment of his physician, as it was known to any one who cared to listen to Hawberk's sanguine prophecies of the outcome.

Mrs. Enderby, lingering honorably out of intelligible eavesdropping, but not out of ear-shot, was disinterestedly impatient of the interval before Anther spoke.

"What do you think," he began, and at the sound of his voice she fled from temptation, "of evil done in the past, and so effectually covered up, except from two or three people, that for the public generally it never existed: should you think it the duty of the two or three, or any one of them, to make it known?"

"I'm not quite sure that I follow you," said the rector, but confessing his interest by his look of prompt animation. He was seeking, as he professed, a stronger light upon it, but he could not feel that Anther cast this light upon it by what he said next.

"Take the case of ——," the doctor resumed, and he named a famous case which once agonized the public with a curiosity still unsatisfied. "He must have known, and a few others must have known quite as well, whether he was guilty or innocent in that business. Do you believe it would have been to the advantage of religion or morals to have had the fact generally known; or was it just as well to have had it hushed up forever, as it apparently was?"

"I don't see what advantage the common knowledge of it would have been," the rector said, still feeling his way rather blindly. "I can't see what use it would have been as concerns this world, to have had the fact known. If the fact would benefit some one, save some one from unjust suspicion, relieve some burdened spirit, yes; but otherwise not, I should say."

"You think the truth itself, merely as truth, has no claim upon our recognition?"

"What is truth?"

"Ah, that's what jesting Pilate asked!"

"Isn't the truth," the rector pursued, "that absolute entirety of fact which includes not only every circumstance, but also every extenuation in motive and temperament?"

"Well?"

"That sort of truth can never be made known in this world, and the brute fact doesn't express it."

"You remand it to the Last Day?"

"I leave it to God. The Searcher of hearts can

alone find it out, and judge it. If we press for judgment here, we are in danger of becoming executioners. But I am never able to deal with abstractions, such as this case has become. You can't lay down any rule that will fit an abstraction. I don't like to lay down any rule at all, except such as I find given us. If there were any particular case—any concrete instance—"

"There is a particular case," the doctor said, "a concrete instance, but I'm afraid that the lapse of time has rendered it as much an abstraction as that other case—in fact, has outlawed it."

The rector could only answer at first, "I should like to hear anything you have to tell me." But he added, "Why are we fencing?"

"Are we fencing? I didn't mean it," the doctor said, with his fagged look and his sad look possessing he rector again with compassion. "I'll lower my point, anyway. I'll go back to the beginning. If a man had so successfully lived what they call a double life that he had kept each life largely a secret from the other, and kept everybody but those he had most wronged altogether out of the secret, and there were but one impartial witness of the facts, would it become the duty of that witness to make the facts known when the man was dead and the evil he did had not apparently lived after him?"

"I think you'll have to be a little more specific."

"Have we no such a thing as a duty to justice? Is there no such thing as justice?"

The rector looked grave. "I have never seen any instance of justice in the world. I have seen many instances of mercy. I should say we have a duty to mercy. We are warned more than once to make sure first of our own sinlessness before we offer to judge the sins of others."

"But imagine that the guilt of the man I am imagining had imposed itself upon the public for virtue, and was apparently left to the Last Judgment, as so many things—most things, in fact, as I agree with you—seem left, and time had gone on till it became, by this chance and by that, the question of recognizing a cruel miscreant as a public benefactor, and holding him up as an example to the young, and celebrating some twopenny munificence of his as an act of characteristic virtue, of habitual greatness and goodness—"

The rector rose, and his face whitened, as the doctor's had reddened with the rush of feeling into his voice. "Are you talking to me of Royal Langbrith?" he asked.

"I am talking to you of Royal Langbrith," Anther replied. "And ever since I heard that you had been asked to take part in this preposterous business I have been talking to you about Royal Langbrith. Not to your knowledge, of course, but in those one-sided colloquies which, I dare say, you hold as well as I when you are working up to face some one whom they concern. When Mrs. Enderby first told me you had been invited by Langbrith's son to join in honoring his father's abominable memory, my impulse was to come at once and tell you what the man had really been. But when that impulse passed, I said to myself that I would think it over;

and I have thought it over and thought it over, but never with so much justification in paltering with my duty as you have given me by the things you have just said. It seemed to me, on one side, that it was an outrage upon your own purity and uprightness to let you go on and unwittingly praise that infamous scoundrel. It seemed an atrocious invasion of your rights, an abuse of your ignorance as well as an insult to your office. Then, on the other hand, I asked myself what harm would be done if I let you go on, compared with the harm I should do if I stopped you—the pain I should needlessly inflict; for the truth would now probably never come out, and in the interest of public morals had much better remain hidden. I recognized this long ago. I saw that the time for a public exposure of the man's evil had apparently passed; that it had paralyzed those who had left it hidden; but when I heard that you had been asked to eulogize such a miscreant in public, I felt a new responsibility. I realized that if I let you do so, I should be guilty towards you; yet, if I spoke, I should be putting my burden upon you, and compelling you to the sophistications with which I stifled my own conscience. You could not then stand up and declare the truth before the people; you could only reveal it to that miserable boy; or, if you had not the heart for that, you must stultify yourself and wound him with lying excuses. I paltered with my duty, and I have come at the eleventh hour to do what I ought to have done at once or never done at all.

Anther told his story with a fulness which he

had wanted even in telling it to Judge Garley. In the sympathy which he felt Enderby was giving him, with that instant self-forgetfulness natural to the born priest, there was invitation which the legal mind could not give him, with its concurrent criticism of his facts and motives. He was dealing now with a man who could appropriate his facts and realize his motives to their remotest intimations and finest significances. Science and religion met in the study of the life laid bare between them. At any detail from which Anther faltered, Enderby prompted him, and, in the end, nothing was left untold.

"Besides Hawberk and Mrs. Langbrith and yourself, is there any one knowing to the facts?"

"John Langbrith; but how intimately he knows them I can't say. We have never exchanged confidences. He was on the train with his brother when Royal Langbrith died. Didn't I say? Yes — he died in the smoking-car coming up from Boston, but so suddenly, so secretly, that John Langbrith did not notice anything till he put his hand on the dead man's shoulder to arouse him from his nap when they reached Saxmills. He had died as secretly as he had lived."

"What has become of the woman?"

"Who ever knows what becomes of the woman? Perhaps, in this case, John Langbrith does. I ought to tell you," Anther added, "that I have put the case to Judge Garley."

"How long ago?"

"Several weeks — a month."

"And knowing the truth, he let me accept a part in this commemoration!"

"You might say the same of me."

"No, I couldn't say the same of you. I can understand the stress there has been upon you, and your reluctance—your fear of being misunderstood—misconstrued. But if Judge Garley had given me a hint— No, I don't blame him either! I mustn't be cowardly."

The rector sat with his elbow on the arm of his chair, and his head propped on his hand, thinking. What he found first to say, with a sigh and a forlorn smile, was, "It's part of my cowardice that I could wish *you* hadn't told me."

"I was obliged to do it. In this, at least, I have had no selfish motive."

"Of course not. But I must go on all the same, you see." Anther said nothing, and Enderby asked, "The boy is without the least suspicion, the slightest surmise?"

"Absolutely. He was not purposely kept so. But the time for telling him never seemed to come. Who could tell him?"

"It may never come," the rector mused aloud, and he said to Anther, "It hasn't come now."

They were silent together, but the doctor spoke first: "It did cross my mind that you might feel authorized to—"

"No," the rector stopped him; "we must leave it all to God now, as it has been left hitherto. He will know when the son can best bear his father's shame. He will know how to do justice, and when, on the memory of the dead; but until now, in mercy to the living, He has forborne. The circumstances will arrange themselves; the atoms will fall into the order of the divine scheme. We must keep our hands off. *De mortuis*—you know the saying; there is as much wisdom as kindness in it. There is a feeling—it is mostly a vengeful feeling, I don't know why—that men's evil deeds must not be suffered to lurk in the dark; but perhaps they should, for this life. What would it avail to have them dragged into the light? Everything shall be made known, but perhaps not on earth. Whoever wished to hasten the knowledge of hidden evil, here and now, might well beware of forcing God's purposes, as we understand or misunderstand them. It could not help this community to know the truth about that wretched man. It would only render it cynical and deprave it. But I am not concerned about the son, primarily, I am afraid; or about our fellow-citizens. I have the selfish concern of keeping myself clear from falsehood in what I have to do. At present, I don't see how I can, but I shall try; and, meantime, between the two evils before me, I will choose that which seems likest virtue."

Anther was struck with the similarity in the conclusions of the priest and those of the judge, but he did not comment on it. Enderby himself offered none of the reflections in which he seemed lost, and Anther, after a little longer stay, in which nothing suggested itself as a solution, took his leave, without protest from the rector. He carried with him, capriciously, the vision of the rector's neatness, as to the black waistcoat, buttoned to his throat, which was without suspicion of those droppings from the rector's full beard such as the doctor remembered noting on the vestments of some clergymen less conscientiously benzined by their wives.

Enderby's wife was otherwise so conscientious that she would not join him in his study, after he returned from seeing Anther to the gate, till he called to her, "Come here, Alice." Then she rustled down-stairs and entered to him with a face eager for the account of his talk with the doctor. At sight of his face, looking up at her from the chair into which he had nervelessly dropped, hers fell.

"Dearest!" she said.

"I am in trouble," he answered. "I want you to help me."

Though a woman whose chief delight was, ordinarily, in the expression and examination of her emotions, she now postponed them, as she was able to do in great emergencies, and closed so promptly and directly with the trouble he owned to her that he was able after an hour to say, "Well, then, I will do it."

"It's the only thing you could do, and it's the thing you *must* do. It's what suggested itself to you at first; and *I* call it an inspiration."

The notion of an inspiration was something left over from her Unitarian nurture, which she would not deny herself in the present exigency. It had a literary rather than a theological significance, and was less an article of faith than of critical appreciation. Then the rector went to bed, and, instead of harassing his worn-out brain by vain dramatizations of the predicament, surprised himself by falling almost immediately asleep.

It was for Dr. Anther to lie awake after he had driven home through the dim Saxmills streets, usually so quiet at half-past ten, but to-night only quiescing, after a tumultuous evening of last details in preparation for the morrow. His course lay by the open square on which the library faced, and he noted that the platform built up around the doorway, below the bas-relief, for the invited guests had been draped, since he passed earlier in the evening, with American flags. The tablet was veiled in white cotton cloth, which, in its association with the dead, gave Anther the sense of a shroud, so that he started at the light, gay laugh which burst from the lips of a girl pausing with a young man and looking up at the platform from the square below. He recognized the voice of Hope Hawberk in the laugh, and in the young man beside her he recognized James Langbrith, and he imagined her teasing him.

He smiled to himself in the prevision of his absence from the group of the invited guests who were to occupy that platform the next day. The committee of arrangements had promptly sent him an invitation; and a second card had come later, under a cover addressed in Langbrith's hand-writing, as if he were not willing that Anther should by any chance be passed over. So far, indeed, Langbrith had subdued his rancor with his old friend. But Anther had determined, from the first, not to be present at the dedication, and he had not faltered since.

The figure of a woman, imaginably some patient who had waited for him in vain, slipped from his gate and went down the obscurity of the street, in the opposite direction, as he drove up to Mrs. Burwell's darkened house. He put his horse and buggy into the barn, and then came round and let himself in at the front door. On the threshold within lay something white, which he felt to be a sealed letter; and, when he had turned up his office lamp, he found it addressed to him, in a hand which he knew. "Dr. Anther," he read, "I want you should not fail to accept James's invitation for to-morrow. He is feeling very anxious you should be there, though he will not say so. If you don't choose to do it for his sake, do it for mine. I would give anything to have you.—AMELIA."

He turned it over, as people turn letters over, rather when they have got everything out of them than when they have not, and he knew that the woman he had seen coming away from his gate was Mrs. Langbrith. Her anxiety must have been great, to bring her from home so far at that hour, and she must have wished to keep her writing him a secret from her household, if she could not send the letter. She might have hoped to see him, and carried the letter to leave in case she should not find him.

"Why," he asked himself, bitterly, "should *we* be doing things by stealth? We hide our affection, as if it were something to be ashamed of. We behave like guilty persons, but you are the most innocent of victims, and I am to blame only for not forcing you to right yourself. I can't stand it, Amelia!"

XXIV

LANGBRITH had at first meant to dedicate his
father's memorial on grand terms. It had seemed
to him not out of scale with the merit of such a man
to have the governor and his staff in full uniform
present at the ceremony. But a few drops of ridicule sprinkled on the notion by Falk extinguished
it, after an angry sputtering; and he reasoned that
to confine the civic interest to Saxmills would be
to intensify it, and to appeal still more strongly to
the local pride. In his illumination, he declined
the offer of even a band from the next town, when
it was submitted to him through the committee of
arrangements, and decided to have no music but
such as the fifes and drums of the Saxmills cadets
could make in their march through the streets.
This, with the singing of the public-school and
Sunday-school children, ranked below the platform
where the invited guests were to be seated, before
and after the unveiling of the tablet, would be taste-
fully sufficient in Langbrith's more tempered ideal
of the affair.

The cadets looked very well as they paraded, and
the children, marshalled by their teachers, looked
charming—the larger boys bearing school banners,
supported by smaller boys holding the tassels on

each side, as they marched to the library and formed themselves in the appointed order. They counted in their number all the children in town, except some inveterate truants in whom the Fourth-of-July excitement was beginning to work, and who opened their celebration at daybreak with the explosion of cannon-crackers. Throughout the morning, the sound of their torpedoes broke upon the more ceremonious sounds of public rejoicing; and, when the procession formed, they made themselves its straggling escort, and followed it in the mixed admiration and derision of boyish outlawry. It had been proposed, at one time, that the mill hands, men and women, should join the procession, in such gala as they chose; but John Langbrith had passively disfavored the plan, which had not found acceptance with the hands themselves. When it was brought to James Langbrith's knowledge, he decided against it, as something perfunctory and out of keeping with the voluntary spirit of the affair.

Falk, who stayed over the week as Langbrith's guest, praised his decision as a stroke of surprising wisdom. He mingled with the operatives, in the rear, where they formed the great mass of the spectators, and was able to report to Langbrith a satisfaction with their unalloyed holiday which he was sure they would not have felt in the procession. He himself refused any share in the ceremonies by refusing a seat among the invited guests; and when he was not going about and feeling the public pulse in Langbrith's interest, he amused himself by making the three young girls under his charge laugh, or try to keep from laughing, at his remarks on the general and personal aspects of the occasion, especially on the activities of Langbrith, as host, and Matthewson, as chief-marshal. Susie Johns was not concerned in either of them, and could laugh at both, without the fond misgiving of Jessamy Colebridge or the perverse delight of Hope Hawberk, as Falk made them note the majesty of Matthewson in ushering the invited guests up the stairs of the platform, and the urbane hospitality of Langbrith in receiving them at the top and appointing them their seats. The girls laughed so much, and Falk kept so grave, that glances of reproval for them and sympathy for him were shot from neighboring eyes, while the band brayed on, and the crowd packed into the square before the library cheered each guest as he mounted and took his place.

They were, first of all, the oldest employés of the Langbrith paper-mills, women as well as men, who were given the seats next the speakers; veterans of the Civil War had the seats behind them; and then the village dignitaries, the selectmen, the high-school principal and the Sunday-school superintendent, with citizens of no official quality, but eminent in business, or entitled to recognition by their age or social standing. Before all sat Judge Garley, Mr. Enderby, Father Cody, the orthodox minister, and John Langbrith. At the last moment, Matthewson was seen receiving Dr. Anther at the foot of the steps, and then Langbrith, with a forward start and a flush of surprise, greeted him at the top. The young man's face was lighted with a joyful smile as he clung to Anther's hand and bubbled an incoherent welcome, looking round to see where he should place his father's old friend. He restrained a movement of Anther towards the rear seats, and led him forward and put him between the judge and the rector, who made room for him with dumb shows of courtesy. The band brayed out afresh, and the general applause of the crowd rose in such personal cheers as: "Hurrah for Dr. Anther!" "Hurrah for Dr. Anther!" Hope took out her handkerchief and waved it, and then Jessamy and Susie took out their handkerchiefs and waved them. The doctor sat down abashed, and his lowered gaze fell upon the veiled face of a woman sitting in the foremost row of chairs, placed in the little square before the library. At sight of Amelia Langbrith, a sad smile overspread Anther's reddened visage, which he turned at the slight tumult caused by some unexpected event at the foot of the steps. The tumult passed with the slow mounting of a figure to the platform, and its momentary hesitation at the top; then the gaunt shape and blotched, brown visage, with the deeply sunken eyes, of Hiram Hawberk showed themselves spectrally to the crowd. Inarticulate cries and gasps broke from it, and shaped themselves in derisions like "Three cheers for Hawberk!" "Hurrah for Hawberk!"

Langbrith turned from whispering to Judge Garley; at sight of Hawberk, he flashed a silencing glance at the crowd, with a scornful lift of his young head, and hurried towards him with outstretched hands. He seized Hawberk's trembling hand as he had seized Anther's, and then, placing it under his arm, led him forward. There was no place among the front seats, but every occupant of them rose and offered his place to Hawberk. Langbrith waved the others down, while he spoke to his uncle. Then John Langbrith, chewing the splinter of wood on which he had been sardonically working his jaws from the first, shook hands with Hawberk, and pulled him into his place, where he took Anther's hand, proffered across their knees, and remained dimly looking out over the people.

"Why, I thought your father wasn't coming, Hope?" Jessamy Colebridge said.

"I suppose he changed his mind," Hope answered, quietly. But she dropped her veil as she rose with the rest at the uplifting of Father Cody's voice in the words of the invocation.

The priest had been chosen for the opening ceremony to satisfy him in certain scruples with reference to his association with the Protestant clergy, which the committee treated with the large indulgence of an underlying indifference in sectarian matters. But afterwards their choice was felt to be almost providential. The dignified form of his words, and the sort of sacerdotal authority with which he pronounced them, struck a note fortunate for the after proceedings, which these obeyed. It did not, indeed, form a law for the excursive generalities of the oration which Judge Garley delivered, but it tempered him to perhaps greater simplicity and directness than he would otherwise have had. He paid a tribute both to the secular and sacred

221

character of the priest, which gathered all Father Cody's parishioners to him, and carried them attentively with him wherever he strayed. But no one followed him so closely, so curiously, as Dr. Anther, who was, as anxiously as he was unwillingly, alert to see what course the legal mind would take among the difficulties so evident to him. It could not be said that Judge Garley made light of the difficulties; lightness was not a thing imaginable of him; but he won his way among them by leaps and bounds, which, if ponderous, certainly got him over the ground, and by turns which, if not agile, were undeniably effective. He made a background of the history of Saxmills, from the earliest colonial period down through the old French War, the Revolution, the last war with Great Britain, and the invasion of Mexico, to the great civil strife for the maintenance of the Union; and then, in the middle distance, he sketched the rise of the manufacturing interests of New England, with their share in the immense expansion of industries throughout the country, after the pacification of the South and the establishment of the great principle of manhood suffrage on the rock of the Constitution. Such, he said, was the time, such the place, such the situation that confronted the man whose far-seeing enterprise had given Saxmills its unsurpassed prosperity, and whose munificence, in one of its many instances, they were tardily recognizing to-day. They all knew *who* the man was; but *what* was he?

Anther held his breath as he watched his old friend standing before the impressive canvas he had prepared, and wondered what manner of heroic effigy he would paint upon it. From where he sat, beside Hawberk, feeling the tremor of his limbs when they chanced to touch his own, and breathing the narcotic odors that exhaled from his person, he could not catch the eye of the orator, which presented itself only in profile, as he shook his head in challenge and pounded the air with his fist in accentuation of his appeal. Except that he knew the judge to have justified himself invulnerably through his professional conscience, he might have thought that he kept his face purposely averted, and he held his breath when Garley resumed. "I never spoke with the man. I never saw him. I never heard his name till I came to live among you here. He was no friend of mine, not even my acquaintance, yet from his work I *know* him." But when Garley reached the end of his characterization of Royal Langbrith, Anther laughed in his heart, with no wish to utter its bitterness to his old friend, as he resoundingly closed with the words: "Such was the man, such was the character, such was the personality whose counterfeit presentment shall be revealed to us this day, and each day shall show him to others after we are dust, as long as stone and bronze shall endure."

On that magnificent background the scenic artist had really painted nothing; nothing but what might pass for one enterprising and successful American as well as another: the mere conventional outline of a face or a figure which a thousand names would fit as well as Royal Langbrith's. He had carefully avoided not only distinctive traits, but he had, with purpose evident enough to Anther, kept a surface as impenetrable as it was shallow. He had given this surface a glare which dazzled the eye and distracted the thought from the performance to the performer; and Anther judged him less and less harshly as he considered that Garley had discharged a duty which he could not shun as harmlessly as it could be discharged. No one but the brother and the widow of Royal Langbrith knew how false an impression he had made; for it could not be said that his narcotized victim realized it, and none save the rector, who was to follow him, knew how false he had been in making it. Anther did not condemn him. Garley, too, was in the grip of that dead hand which seemed to clutch every one by the throat, and his severest feeling towards him was for the deceit which he had practised upon the son of Royal Langbrith. He could see James Langbrith, where he had retired from the platform to the place beside his mother, watching the speaker with what Anther felt a piteous intensity, and hanging upon every empty word. With tender compassion Anther wondered if he felt the hollowness of the tribute paid to his father's memory. He was touched for that poor, generous boy, and ashamed more than he was amused for his old friend in the success of his fraud. When the applause swept the orator to his seat, and then refluently bore him, bowing and smiling, back to the front of the platform, the young fellow started forward, and, all glowing with tears and smiles, stretched his hand up to the judge, and the judge stooped down to take it. Anther dropped his eyes and hung his head, and he had not the courage to look up again till he heard Enderby beginning, very gravely and measuredly, the address which the dead man was requiring of him, in his turn. Then Anther's pity was no longer for the trusting boy, but for the good man, compelled to this office, and he wondered how he would reconcile it with his conscience.

Enderby stood clutching the scant lapels of his clerical coat, and looking pale above its black. He said that he had been asked to speak some words concerning the ethical significance of the business they were about, and he would now only suggest a few general notions in regard to the respective attitudes of giver and receiver, in the matter of public benefactions. Such benefactions were likely to be more frequent in the future than in the past, when the town had become debtor to one of its citizens for this library, the most useful of its possessions, and the most sacred, after the houses of God; and they must be more and more impersonally regarded. The town was here in its collective capacity to make acknowledgment of the gift, tardily, it was true, but not the less gratefully; for, in the years that had elapsed, the people of Saxmills had fully experienced the great advantage bestowed upon them. It might, perhaps, have been wished—it might, perhaps, have been more graceful in some aspects if the town itself had offered the memorial it was accepting; but in that case it would have anticipated the act by which the son renewed, as it were, and confirmed the father's deed. For himself, the rector said, he was more interested in this renewal and confirmation of it, than in the fact of the original gift. It spoke well for the young man whom in different ways they all knew, that he wished to testify his reverence for his father's memory by doing again one of the best things that his father could have done. In this he had not only testified his reverence to his father's memory, but had borne important witness to the imperishable vitality of a good deed in this world. It was not only the evil that men do which lived after them, but the good also lived, laying upon the future a more powerful obligation to virtue than any bond to vice that evil could impose. God had apparently willed that the good should continually and eternally show itself, and the evil should hide itself, for evil, brought into the light of day, corrupted, and good, whenever manifest, purified and restored and strengthened all men for good. Such, in fact, was the potency of a good deed that, if done from the most selfish motive, it took no color from the motive. It returned through its beneficent effect upon the world to the God of goodness. But they who were assembled to receive from the son the evidence that he renewed and confirmed his father's gift to them had really nothing to do with the character of either. They had only to do with the good-will expressed in what was now their joint gift, and they were to honor neither of those men, but only their good deed, which was not of them, but of God. Few present had known the elder of the two; all present had known the younger, and it was he who stood for both before them. Every heart must respond to the impulse which had governed their fellow-townsman in his filial devotion to his father's memory, and must rejoice with him in the beauty and fitness of the tribute he had paid it. If either were to be known by the other, though it was not necessary, for the present purpose, that either should be known apart from his gift, let the father be inferred from the son, and let them not be separated in the public acceptance of their benefaction.

The rector would have sat down; but James Langbrith, who had remained on the platform after Judge Garley's oration, prevented him. He seized Enderby's hand, and Anther heard him say, while he clung to it, "You have spoken just as I feel my father would have wished you to speak. He was the most reserved, the most impersonal of men, and I thank you, thank you, thank you for him as well as myself."

"Oh," the rector groaned, in a sort of protest; but before he could say anything the leading selectman rose in his place and commanded, "Three cheers for both the Langbriths!" James Langbrith stepped forward to acknowledge the applause, and Anther felt Enderby's eye seek his own.

There was no defiance in the rector's asking look, but a sort of entreaty, as if for the effect his words might have had with the man who knew how, primarily, they had been spoken to him. Enderby's back had been turned to Anther while he addressed them to the people, but it had not needed the comment of the speaker's face to convey all their latent meaning to Anther, whose eyes were as troubled as his own. He put out his hand and sadly pressed the hand of the rector, who miserably smiled a little.

"You did the best that any man could, in the circumstances," Anther said, under cover of the uproar.

"Now, friends," said the selectman to the crowd, when the cheering had died away, "the tablet will be unveiled."

At the moment James Langbrith stepped back to perform the office, Anther saw Hawberk put something into his mouth and heard him huskily explain, "Thought I might need some, and brought along a little of the gum."

Langbrith pulled at the cord which had been contrived to separate the white curtains veiling the tablet, and slip them to the sides on the wire from which they hung. The contrivance would not work, though he tugged and twitched, and there began to be some nervous laughing in the crowd, which had its effect with him. He gave an impatient pull and the whole contrivance came away, dropping to the ground behind the platform. A girl's hysterical cry went up, and the people began to clap and cheer. Langbrith had turned an angry face towards them, but their good-will was so manifest, their laughing had been clearly so helpless from the sense of humor which any unserious mischance appeals to in a crowd, that the anger went out of his face, and he, too, was smiling when the voice of the selectman announcing that the Rev. Mr. Alway would ask a blessing recalled him to the necessity of a more appropriate expression.

While the people were stirring vaguely from the attitude in which the benediction had left them, Langbrith came forward and shouted, "Friends, ladies and gentlemen, there's a lunch at my mother's, and everybody is invited—everybody!"

The crowd cheered and the band played and the square emptied itself in the direction of the Langbrith homestead.

XXV

The last of the guests had got themselves away
from the Langbrith grounds late in the afternoon,
with the difficulty that people unaccustomed to
social rites find in taking their leave. It was half-
past four o'clock when Langbrith stood, with his
mother, in the porch at their front door, looking
down, over the trampled lawn and dishevelled decorations, at that fellow-citizen who managed all the
public functions of Saxmills, rushing about in his
shirt-sleeves and directing his shirt-sleeved helpers
in the work of dismantling and removing the long
tables of rough board at which the hungry throng
had lately joked and shouted and rioted.

The son noted the knot between his mother's
eyes, and laughed. "You'd like to go out there
and take a hand, mother," he interpreted; "but
you'd better leave it to Danning. It 'll suit him
better." He sighed deeply. "It's been perfect,
mother, beyond my dreams. It's been beautiful,
ideal. I couldn't tell you *now*, without disturbing
my sense of it, how happy it's made me. It's made
me feel as if the people here loved me, and I do like
to be liked, though I don't know how to show it,
and that they cherish my father's memory. How
good everybody has been—how kind! It was aw-

fully sweet of the old doctor to come and sit on the platform after his reluctance. I won't forget it." Langbrith gave a short laugh. "He knew father better than I do, and he probably felt for him against the affair; but if father had cared to look down on it to-day, I can fancy his being pleased with it in some shy, reticent way. I wish the doctor could have come to the lunch."

"He said he had a patient—over at Wakeford," Mrs. Langbrith said. "I asked him to come."

"Yes, I know. I hoped he might have got back. Well, now, you must go in and lie down, mother. Take a good rest." He put his arm round her waist and pressed her in-doors, and got his hat in the hall. "I'm going to pick up poor old Falk somewhere. I shall probably find him at the Johns', unless Jessamy got away with him."

He kissed his mother and left her, not to lie down, but to go and take counsel with Norah about the things that Danning's men would be bringing in to be washed up and put away. He saved his conscience with respect to Falk by walking past the Johns', and looking in over the fence, but he did not stay to ask for his friend on his way up to the Hawberks'. He did not know whether he had seen Falk sitting with Susie Johns at her door or not. Every sense of his was full of Hope Hawberk. Except as she was related to them, she pressed even the facts of this happy day out of his consciousness.

Hope's grandmother came to the door, and said with grim directness, before he had asked, "She's round in the garden."

"Oh!" Langbrith answered, and he took the little path in the grass that the feet of the household had traced round the corner of the house.

Hope was sitting in a low rocking-chair, by the dial, which the sun had relieved from duty for the day by getting down among the tops of the pines on the hill. She was reading a newspaper, but she was not so absorbed in it that she did not hear his step sweeping over the grass. As she looked up she laughed quietly, and in her laugh he felt a peculiar note of welcome. "Well, how did it go off?" she asked, and she let fall her paper and rocked back in her chair.

"Don't let's talk of it," he said, and he crouched at her feet, with his back against the base of the dial. "Let's talk of ourselves."

"Well, what about you?"

"Nothing about me. When I say ourselves, I mean you, for you are ourselves. At least I am nobody without you."

She laughed again, but her derision was full of the love which she did not try to keep out of her eyes. His own eyes glowed upon her. Neither felt the need of speaking till she turned her head away with a little difficult motion, almost as if it hurt.

"Then you will?" he murmured from somewhere deep in his throat, and she answered, low:

"Yes."

He bent forward and put his head on her knee.

"Don't be silly," she said, with a catching of the breath, while she smoothed his hair with her hand.

There was no other demonstration between them,

232

because he knew that she liked best that there should be none, and it was a moment before he lifted his head, with a laugh of the joy otherwise unutterable: "I knew you would say yes, now. But why now? Why never before?"

She looked at him with the glowing eyes which she could not keep from his face, but it seemed to him that she no longer saw him so distinctly, for a mist that veiled their glow. Her lips twitched so that she could scarcely form the words: "Can't you think?"

"No. What have I done?"

"You want to make me tell you! How you acted to father—when—when they laughed—I said that I would do anything for you, then; I said that I would do anything you asked—"

"Hope!"

"Don't make me cry! I shall hate you if you do. When I need all the strength I have, so!"

"No, Hope; but listen to me. I must be honest. I didn't do that for *you*. I did it for *him*. I like your father; he was my father's friend; and I had nothing in my mind but the thought of their old friendship. That needn't make you cry, or, if it does, it needn't weaken you. Hope"—he kept getting her name in as often as he could, for the pleasure of speaking it—"I am not going to ask any promises of you, now. We will let the future take care of itself. But I want to tell you; I haven't told my mother yet; I am going to Paris to study—to study the stage, and learn to write for it; I believe I can write plays, and Paris is the place to study the stage.

I thought I should ask you to go with me; but I see I can't"—she shook her head in affirmation of his words—"but if I can take your love and leave you mine, will you—will you—wait?"

"Yes."

"Oh, Hope!" he sighed.

"Oh, James!" she sweetly mocked him.

"Where was I?"

"You had left me waiting."

"Well, that is all, then."

They both laughed.

"Of course," he took up the broken thread, "I shall tell mother."

"You couldn't go without."

"Oh, I mean about you. She will be glad. She likes you so much, Hope."

"Well, I like her, too."

"And you will go to see her often, Hope, won't you?"

"Not often enough to cause remark," she drolled, and he laughed and said:

"How funny you are, Hope! Falk thinks you are the wittiest girl he ever saw."

"Well, you've always told me Mr. Falk hadn't been in society a great deal. There must be lots of funny girls in Boston."

Langbrith thought that droll, too. "I believe I love you more for your fun than your beauty, Hope."

"Perhaps there's more *of* the fun."

"No, I don't say that. You are the most beautiful creature in the world to me. And Falk thinks that your dark style—"

"Well, I always thought Mr. Falk was pretty, too. So it's an equal thing. Now, we won't talk of that any more; it's too personal. We will talk about Paris. I shall never dare to tell grandmother that you are going to write plays. She thinks I'm bad enough as it is, and if she knew that I was engaged to a person who wrote plays, she would certainly give me up. Does Mr. Falk know about your plan?"

"Why, he's going with me! Hope! May I tell you a secret?"

"Well, if it isn't a very large one."

"It's nothing. You know he is going to be an artist, and Paris is the place for art as well as the stage, and I am going to lend him the money. I'd give it to him, if he'd let me. What better use could I make of it? But of course Falk won't stand that."

"No, I wouldn't, in his place."

"Does he care for—I mean does Susie Johns care for him?"

"She never said so. Perhaps she hadn't been asked. She's rather queer, that way. She never answers till she's been asked. She's very secretive."

He laughed, and began in another place. "I wish I could have you with me to keep me from playing the fool."

"Why, I'm the greatest fool myself," she explained.

"No, you're not. You're the very soul of common-sense. But I shall keep writing to you, and consulting you about everything, and that will make me sensible. And perhaps—in about a year—"

She mocked, "I was just waiting to know how long!"

"Hope," he asked at another tangent, "Dr. Anther *does* think your father's getting better, doesn't he?"

"He thinks his will is getting stronger."

"I understand you can't leave him, Hope, and that's why I don't ask *you* to go with me to Paris as well as Falk; but when your father is all right—and he *will* be, I *know* he will—then we will go out together—my mother and your father, as well as you."

"What a beautiful vision! And what about grandmother?"

"Oh, we would take her, too."

"I should like to see you getting grandmother on a steamer! Why, she thinks going on the cars is as much as her life's worth."

"We can manage, somehow." They laughed together at his optimism, and he asked, "Do you know what I liked best in the whole thing to-day? I mean besides your father's coming. Dr. Anther's being there. He didn't like the notion of the tablet at the first, and he let me feel it; but it was just his way—working round, and giving in handsomely in the end, without saying anything. My heart was in my mouth till he came onto the platform. It wouldn't have been anything without him."

"Of course it wouldn't. But, of course, he was sure to come. He's grand."

"Yes, after my own father, as I imagine him, there's nobody equal to Dr. Anther, as I know him."

They talked rapturously away from themselves, and they talked back in ecstatic return, and an hour passed before he reverted to her with impatience of anything but her in her relation to himself. "What made you cry out that way?"

"Me? How did you know who it was?"

"Don't you suppose I should know your voice, in the dark, anywhere in or out of the world? What made you do it?"

"As if you didn't know! I was so worked up by those curtains not coming apart, and thinking how you felt, that I couldn't help it, though I wasn't sure but it was somebody else. If it had gone on much longer, I should have got onto the platform and married you on the spot."

Langbrith jumped alertly to his feet, and Hope rose, too, laughing. He put out his arms towards her. "Now I think it's full time for you—"

She did not try to escape, but a sound of lamentable groaning came between them, and she called out, "Oh, poor father!" and whirled from her lover into the house.

He stood dazed by the ghastly interruption, and remained bewildered when, a little after, she returned to him, somewhat paler, but not looking as distressed as he looked, and dropped again into her chair.

"Isn't there anything I can do? Go for Dr. Anther?—"

"No, no! It's all right, now. He was just dreaming—he has awful dreams, but they are only dreams."

"Oh, Hope!" He stood before her, not offering to take his place at her feet again, but aching, as she saw, with pity for her.

"You mustn't mind me. I'm used to it. And it isn't anything real, you know."

"It seems terrible. I don't know how to bear it for you."

Hope smiled. "Well, you don't have to, and I can bear it for myself as long as—as long as father must bear it. Are you going away?"

"Yes, I must go back to mother—"

She rose, and, without his advance, put her arms round his neck and kissed him, and then began to cry against his cheek. It was not the passionate embrace with which he had often, in his burning reveries, sealed their betrothal, but it was something sacreder, sweeter, and he seemed purified and uplifted, as if her arms were raising him into heavenlier air. He knew now what misery and sorrow, what squalor, even, he was making his part; but he thought only of her with whom they came, and he was richly content.

"Your trouble shall be my trouble, after this," he began, but she would not let him say more.

"Yes, yes! Don't talk!" and while she brushed the tears from her eyes with her handkerchief, she pushed him from her with the other hand.

He accepted his dismissal. "I shall come back after supper," he said, and she neither invited nor forbade him. He did not go home; he could not, without first using the new authority which her love had given him, and he went round by Dr. An-

238

ther's office to ask him if nothing, nothing, could be done for her father. He tried to think about it all, and how he should press the doctor to some conclusion, to some definite promise, to some clear prophecy of a fortunate end; but it was confused in his mind with his love, and he was so lost in the sense of that as it concerned her and him alone that from step to step he forgot what he was about and had to recall himself to his errand. Once he went down a wrong turning, and, when he came to Mrs. Burwell's at last, he recognized the house with a kind of astonishment.

"The doctor ain't here," Mrs. Burwell called down to him from the window over the door as he stood with his hand on the bell-pull. She had her head tied up in a handkerchief, as if she had been sweeping; the impression of this was strengthened by her having a broom in the hand that supported her on the window-sill. "He hain't got back from that patient—drefful sick crittur, I guess—to Wakeford, and I'm givin' the place one last dustin'; I don't know when it 'll get another. I was *ril* sorry I couldn't come to the ceremony to-day, but I got my mind set on finishin' my movin', and nothing couldn't seem to stop me. I feel bad about leavin' the doctor here, alone like a cat in a strange garret, as you may say, but I guess I got to. I don't know who he'll get in to care for him. As far forth as I can make out, he ha'n't even thought of anybody."

"He'll be in after supper, I suppose?" Langbrith said, with an imperfect sense of the words spilled on him, as in a stream, from above.

"Yes, if he *gets* any supper," Mrs. Burwell responded with mystery lost upon Langbrith's abstraction. "He's always in nights, you know, without he's got a call."

"Then I'll come round again, later."

"So do!" Mrs. Burwell called after his averted figure as he stepped down the two yards of path to the gate, and moved away with feet that wandered with his wandering thoughts.

Something had penetrated the whirl of his mind which centred around the idea of Hope all kindly and pleasant things, and he was afterwards aware of some meaning in Mrs. Burwell as to Dr. Anther which he had not taken at once from her words. Had she meant that the doctor had bought her house or hired it? He had lived there a long time, and it might very well be. But a magnificent scheme now suggested itself to Langbrith, which he would consult his mother about, and then propose to the doctor, if she approved. He would offer Dr. Anther his father's office, standing apart from the mansion, if he found he had not taken Mrs. Burwell's house; it would be more convenient for him, and it would be near the hotel, where the poor old fellow could get a meal at any time without being subject to such severities as Mrs. Burwell had practised with him, and as he must fall under again if any village person took him to board. Langbrith himself would feel so safe, having him there near his mother, for all advising and helping in any sort of exigency. With that lifelong friend near her, he would not feel as if he were leaving her alone for the year he should spend in Paris, before he brought Hope home to the old place.

He glowed with the thought of what motherliness and daughterliness there would be between those dearest women, and how he would protect and cherish them both in their common reliance upon him. He wished Falk was there. He would like first to consult Falk about it. Falk had so much sense, and would put his finger on any weak spot in the plan and laugh him out of it if it would not do. He felt the need of Falk so much and the desire of immediate action so greatly, that he turned from going home and walked rapidly up the hill towards Susie Johns'. He wished he could go and ask Hope's counsel, too, but it would be silly—he feared her thinking it silly—if he went back to her so soon; and if Falk approved, he knew that she would, and his perfected plan would be such a pleasant surprise for her.

He could make an excuse with Susie Johns, that he had come to fetch Falk home for tea; but, when he knocked at her door, the Irish girl who answered him said that Mr. Falk was at tea within.

"Oh, then, don't bother him," he said, and got quickly away, lest Susie should run out and hospitably seize upon him for another guest. "Don't say who it was," he called over his shoulder to the Irish girl, as he fled.

It would only be postponing the matter a little while. He could see Falk before he saw the doctor, which would be before he saw Hope again, and, with the affair settled in his mind, he pushed down the side-hill street up which his own house looked. He had not reached the bottom when he foreboded a temptation beyond his strength at sight of the doctor's shabby old buggy and his sleepy horse slumped before the gate. But now he suddenly recurred to the thought of bringing him to book about Hope's father, and getting his mother help to get something like a promise of Hawberk's recovery from him. He fancied first telling his mother and their old friend together of his authority for anxiety in the matter. Both these things must come before the offer which he wished to make, and which he now knew he should make without asking Falk about it. But which of the two pleasant things in his mind should come out first was the happy question with him as he entered the wide-open front door and pushed into the twilighted parlor.

XXVI

ANTHER had driven home from Wakeford with a heart softened more and more towards what had been the odious self-compulsion of the day, by his thoughts of the pleasure that had shone upon him from James Langbrith's face when they met upon the platform before the veiled effigy of Royal Langbrith. There had been a fantastic moment when it seemed to him as if the father's misdeeds might be uncovered when the son tore those curtains from his face, but nothing had been revealed; and all the fortuities—one could not quite call them providences—had joined to keep his evil life still in the dark. Anther submitted; he had said to himself he could do no more than he had done; he was not sure that he had done unselfishly in the business, so far as he had acted, and yet he could not have done other than he did. That was his consolation; and now he was going to let events drift as they would; he would never again attempt to stay or steer their course. He had even meant to come to the luncheon at the Langbrith house, and though he had gladly spared himself at the call from Wakeford that reached him when he left the platform, yet he was coming now to make his excuses to the young man for having been unable to take a further part in the affairs of the day.

He found Langbrith's mother alone when he went in at the door, on which he tapped with his whip-handle, and then entered without staying to ring. "James not here?" he promptly suggested in sitting down before her, with his hat on his knee; he waved her away when she offered, mechanically, to take it.

"No," she answered, "he said he was going out to look for Mr. Falk. Perhaps he went to Hope's, too."

She let her eyes fall and sighed "Yes" when Anther said, "It would be the best thing," knowing that he meant as the only atonement the son could make for the father's wrong. "He has always liked her," she added, "but sometimes I have wondered whether she liked him. She's a strange girl."

Anther said, suddenly turning from his wish to let things drift to something in his immediate thought, "And there is the question of how she would feel towards him if she found out, some time, the sort of injury she had suffered from his father through hers."

"Surely she wouldn't hold James responsible for that!" Mrs. Langbrith started as with a physical pang. "How will it ever come out *now?*"

"I don't know. If Hawberk—"

"What?"

He did not answer, but, "Amelia," he asked, with a compassionate intelligence for her helplessness, "why do you cling to this hope of concealment? We have let that poor boy go on and stultify himself, and involve, innocently enough on his part, two good men like Garley and Enderby in the fraud that he has practised on the community—"

244

"Do they know?"

"I had to tell them." She caught her breath, but did not interrupt him. "That's all nothing, though, in my regard, compared with the harm you are doing yourself and the trouble you are storing up for the future, when he finds it out, as he must some day, and asks you if you had known it all along. What will you say to him? I wish you would tell him now, my dear, as soon as you see him, without an instant's delay—"

"I can't, Dr. Anther; it's too late. I can never tell him now."

"Then let me!" It was always coming to that with him.

"No, that would be worse. What would he think of *your* concealment —your being there to-day. But I made you!"

"Yes," Anther sadly owned, "I was there because you asked it. I would certainly never have dreamt of being there otherwise." He rose.

She rose, too, and wavered towards him. "Don't you think I knew you did it for me? Don't you think I felt it? And James," she added, incoherently, "he felt it, too. He cared more for your being there than for anything else, he said." Anther laughed forlornly. "Oh, don't despise me! I know I'm a coward, but don't *you* despise me, or I shall die!"

"Despise you! There's nothing but love for you in my heart, Amelia. Why can't we be all in all to each other?"

"Well," she answered, abruptly, desperately, "I

will do what you ask. Now I don't care what happens. I care more for you than for all the world. Don't you know that?" She stole her arm tenderly through his arm, and pulled herself towards him, but almost at the moment he saw the fondness die out of her face and her arm slipped from his.

He turned and confronted James Langbrith standing in the door-way and staring at them. It was his impulse, somehow, to put himself between the mother and the son, but a guiltless shame withheld him and silenced him when he tried to speak. He heard Mrs. Langbrith gasping, " James, I want to tell you that Dr. Anther—that I—that we—we are going to get married," and he realized that, in anticipating him, she was heroically acting on her instinct as woman and mother.

"Married!" Langbrith echoed, and now he looked at Anther alone, as if for explanation of something unintelligible and incredible. He smiled faintly, and Anther replied with a sudden resentment.

"Yes, I have been attached to your mother for a long time. She has known it, and has consented to marry me."

The resentment was for his own shame, rather than for the young man's words; but not the less it kindled the cold amaze in Langbrith to a flame of hostility that lighted up the whole past of conjecture and misgiving. As one thing after another grew clear in this illumination, the young man's anger burned within him, not so much for the fact immediately before him, as for the series of facts by which he had been duped. But curiously concurrent with his swift retrospect was the flow of his tenderness for Hope, his sense of her love for him and of his love for her, so that it was partly lost in this, and half incredulous, that he began:

"Have you kept it from me so that you can crown my father's commemoration services with it? Was it a surprise you were holding back for me, or were you afraid of telling me?" His anger gained somewhat upon his love, through the mere utterance of the offensive words, but he did not yet speak with a single mind. What was this case, and how did his father enter? He had that still to work out in an unalloyed consciousness.

"Afraid!" Anther dropped Mrs. Langbrith's hand, which he had caught up, and started forward, but he stopped at her cry of "Justin!"

"James," she implored her son in turn, "you don't know what you are saying! Yes—we *were* afraid. I wanted to spare you—I wanted to wait—"

Now he answered more definitely: "And this is your notion of sparing me! Did you choose this time, of all others, to tell me that you had forgotten my father?"

"Oh, you don't know him. You don't know what you're saying. Indeed—"

"The trouble is that I don't know what *you're* saying. I can't make it out. Is it some wretched joke? Dr. Anther, you know how I have always felt about my father. If you were in my place what should you say to a man in yours? It must be distasteful to any son for his mother to marry again, but perhaps you have special reasons that would reconcile me."

His words were temperate, but Anther felt the bitterness that they covered, and he answered as caustically. "I think I could give you special reasons," he said; but at Mrs. Langbrith's imploring look he stopped.

Langbrith had missed the look and its significance. With the sense of Hope fading more and more, he was able to say: "I can imagine them. It isn't the first time that I've suspected you of secretly hating my father, with some such just cause as a nature like his could give a treacherous nature like yours!" He knew, somehow, that he was hurting Anther less than he was hurting his mother, and less than he was hurting himself, even. His rhetoric rang false to him. He was aware that it did not apply. He forced the added words: "But I don't care to know your reasons. I have done with you, sir. I don't want to hear anything more from *you*." He turned from Anther arrogantly. "Mother, what was it *you* were saying about my father?"

She found Anther's hand again and clung to it. She only said, "I'm going to marry Dr. Anther."

"Is that what you have to say about my father? Well, perhaps it is enough."

"Dr. Anther is the best friend I've ever had in the world, and—" she hesitated. Langbrith stood silent, his mind whirling from point to point without seizing definitely upon any. His mother ended, "He will be a good father to you, James."

At this feeble conclusion, Langbrith's daze broke in cruel sneering. "I am of age, and I need no father but the one that I have lost, and that you have forgotten."

"I haven't forgotten him," his mother answered, with a struggle for courage; "I'm remembering him now as I never did before."

"I don't understand this," said Langbrith, haughtily. "But it doesn't matter. I begin to understand some other things, though. I see now why this man has taken the part he has towards my father's memory, but why he should have had the base hypocrisy to-day—"

"He was there because I asked him," she interposed.

"No matter why he was there; his presence was an insult to the living and the dead, and as this happens to be my house, my father's house, I object to his remaining in it another instant."

"James!"

Anther's hand shook in Mrs. Langbrith's clutch, and he burst out: "How dare you talk so to me! If it wasn't that you don't know what you're saying —if your ignorance wasn't so monstrous— But I can tell you—"

"Oh, Dr. Anther!" Mrs. Langbrith implored him, and he stopped, panting. "Will you listen to me, James?" She turned to her son.

"Yes, mother, as much as you like. You can't say anything that will change me towards this horrible business, but I will listen."

"Oh, you don't know *what* I could say to you!"

she broke out. But then she turned again helplessly to Anther. "Will you—"

"No, you must excuse me there, mother; I could
not hear anything from a stranger about family
affairs."

"Dr. Anther is *my* family now," she began,
bravely.

"That is what saves him from the only answer a
gentleman could make to his impudence."

She felt Anther's arm grow rigid under the hold
she had laid on it. "Well," she said, with a helpless
pathos, "as my son will not let my husband speak
for me, I will go with my husband and not speak."

"No, Amelia," Anther said, with the dignity he
had lost in his angry burst; "I will go, and you can
say what you wish to your son."

"I will say it before you or not at all, and if you
leave this house I will leave it with you. I'm not
going to justify myself to you, James." She turned
to her son. "I need no justification—"

"I am not requiring you to say anything, mother."

"And you won't hear me then, my son?"

"If you have no need of being heard, as you say,
why should I put you to the trouble of explaining
anything? I ask no explanation now. It seems
that I've been living all my life in a mistake. That's
all. I supposed we had the same ideal, and that
the memory of my father was as sacred to you as
to me; but it wasn't, and that's your sufficient justification."

"Amelia," Anther entreated, "let me leave you
with James."

"Not for a moment!" she returned. "I can't stay without you, now."

"Perhaps we can simplify the situation by my leaving you with him," Langbrith said. "As it is not convenient for you to let me have my house on my own terms, I will go to the hotel. I can find Falk and go to Boston. When I come back, I hope I can have my house to myself." He recalled himself to add, "You will always be welcome in it, mother."

He turned and went out and left them standing there looking at each other.

"Why didn't I speak? Why didn't I tell him?" Mrs. Langbrith was the first to break their silence.

"I saw you try. It was too late; we're always saying that. Amelia, if this trial is too great for you, I shall never blame you. It has been all sudden and unexpected; no one thing more than another. I didn't dream of your consenting when I came here. Give me up, if you will—"

"And be left with James? Oh no! I care more for you now; perhaps I always did. He was always hard. It seems a strange thing for a mother to say of her son, but it's true; and now he has been cruel. It's worse even than I thought it would be. I'm afraid of him!"

Anther felt within him a curious shifting of the grounds of judgment, and he spoke from the change. "You mustn't condemn him. You must remember how much he had to bear; thinking of his father, as he did, it must seem like sacrilege to him."

"Unless he could know the truth. And if it's

too late for the truth now, take me away from the lie. I can't bear it any longer. Can't we live somewhere else?"

He took her literally, and her shapeless longings for escape crystallized as he answered, simply, "I've bought the house where I've lived."

"Oh, *have* you?" she cried, with hysterical joy. "Then take me there. Let us go now — this instant."

"To-morrow. We can't go now, you know, Amelia."

"I forgot. Now you see how long I have seemed to be married to you. Do you like that? I wish I were! I can't endure to pass another night under this roof! It's hateful! hateful! hateful!"

"Well, you must have patience. You must part kindly with your son."

"With my son? With Royal Langbrith's son?" Her bitterness expressed to him all the revolt of her soul from its long slavery.

He rose in his self-control over her headlong impulse. "You must try to be friends with your son. Nothing else will do, Amelia. If he comes back here, tell him we are to be married to-morrow. Ask him to be with us. You have hidden so much from the world so long that you can hide this, too. We mustn't make our marriage a seven days' wonder. You will feel differently towards James. I pity him from my heart."

"I don't."

"You will, and you must do your best to be reconciled with him. I want your life to be free and happy, from this on. I can't let you incur any shadow of self-reproach. You mustn't have one regret to chain you to the past. Good-night, my dear. I must leave you here because there's nowhere else. But when James comes back you will see him, and try—for my sake—to make peace with him. Remember that his error is not his fault!"

"It is my fault."

"It is no one's. I can understand—and tell him that I beg his pardon for not considering at once—what a bolt out of the blue this has been for him. We have known for a long time that we should marry, but he has never imagined it, and it seems a wrong to his father, as he has idealized him. He can't help acting as he has done towards us, but he will learn to act differently. Yes, his common-sense—and he has plenty of it in the end—will teach him that we could have meant no wrong to his father if he were the best of men. Don't let yourself be tempted, now, to tell him the truth. It could do no good: only harm. Be patient with him. Bear everything from him. He is deeply hurt in the part that is the best part of him; think of that. Amelia, ask him to be present at our marriage. You asked me to be present at—"

"Yes, yes, I will. I don't care what he says to me!"

"That's right. I'll have Mr. Alway. It needn't be in the church, then, it can be—"

"Here?" she shrieked. "In this house?"

"No, in the minister's; and good-night again."

253

XXVII

In the quarrel which he had forced with his mother and Dr. Anther, Langbrith was sensible throughout of failing to say the worst. He had not put into words the outrage which was burning in his heart.

He had not expressed the amaze, and far less the abhorrence, which he felt. He had meant to hurt Anther to death, so far as insult could kill; and he had meant to wither his mother with shame. But the cruelest blows he dealt them had seemed to fall like blows dealt in nightmare, as if they were dealt with balls of cotton or of down; and he had left them in possession of the place he ought to have driven them out of with ignominy.

He was aware of having been disabled for his part by the confusion which still kept him from a clear sense of what had befallen, and perhaps saved him from its full effect. He had entered upon that scene with his soul full of the good-will, the tender purpose towards Anther, which his happy love for Hope had inspired; and he had not, even yet, after all that had passed, wholly freed himself from it. He kept recurring to it with puzzle and interrogation, as something which in its strange metamorphosis he could not make out. It was still mixed with his thought of Hope. It seemed as if he were going to tell her of it still, as he had meant to do, and to taste the pleasure of her praise for it.

He could not make definitely out what he was now really going to do; but he acted upon the notion that he wished to find Falk and get him to take the train with him for Boston. He was sure that he wished to get as far away as he could. That was the first thing. The next thing was to get away from the humiliation of failing to do justice to himself and his cause. Now he saw a thousand proofs that the offence done him had been long impending; that if he had not been a fool, and blind, he must have known it; but the longer it had been impending, the greater the shame, the greater the defamation, the viler the insult to his father, to have it follow so instantly upon the consecration of his memory. His heart closed about the thought of his father with an indignant tenderness, which, somehow, could not leave his mother out. She had always been part of that thought, and he had an impulse to entreat her against herself, as if being a child she had struck him, and there was no one but her for him to go to for comfort.

His feet set themselves uncertainly, as if his vertigo were physical, while he pushed on, looking crazily for Falk. He could not go to Hope yet.

The Irish girl answered him, at the house which he had left so gayly such a little time before, that Falk had gone out to walk with Miss Susie. He asked "Where?" but the girl could not tell him, and he realized that he must not try to follow them. He could not go home, and he would not see Hope.

But he could pass her house; there was that left him to do in the wild need he had of doing something.

She was at her gate, waiting for him, as he knew, after he had bridged the hour of his absence with a recollection of the promise to come back which he had given her. The full moon was looking over the eastern shoulder of the hill, behind the house, into his face; but it was with an inner sense, the vision which love so soon supplies to women, that she read something strange in it.

"Why, James!" she said.

"Come with me, Hope," he bade her, and, as she joined him, wonderingly, letting him seize her hand and pull it under his arm, as he pushed away from the house, up the road climbing into the shelter of the pine woods beyond, "I've got something to tell you, Hope; something to—tell—you," he forced himself. "My mother is going to marry Dr. Anther."

"How glorious!" she shouted, pulling her hand out of his to leave herself the freer to front him.

"Glorious?" he faltered back.

"Yes! I have always thought what a splendid thing it would be. They are such old friends, and they are just suited to each other: your mother is the best woman, and I think Dr. Anther is the best man, in the world. Yes, it's what *I* call glorious."

"*I* call it infamous!" he said, in a voice that struck her with greater amaze than his words by its dreadfulness.

"Why, James Langbrith!" she gasped.

"Infamous. Does no one," he demanded, turning his severity upon her, "remember my father?"

"Why, yes—yes, of course—"

"Is it glorious for my mother to forget him?
Could *you* forget *me?*"

"No, never! And I don't believe she's forgotten
him. But it's a different thing from you and me.
She knows that you will be leaving her some day—
why, I intend to take you from her myself, and, if I
could do such a thing, what mightn't others do!—
and then she will be alone; and why in the world
shouldn't she marry such an *old friend* as Dr. Anther? It would be different if it were a stranger,
and I shouldn't blame you, then, if you were morbid
about it. But Dr. Anther! Why, he's always
seemed to me like one of the family. Why, it's
ridiculous! What has it got to do with remembering your father? Now, James, if you let yourself
get to thinking this way about things, I shall be
afraid to marry you. I say it's the best thing that
could happen, and I can't understand you."

"No, it seems you can't."

"Oh, very well!"

"I don't mean that," he made haste to save himself. "No one can understand how I have always
felt towards my father. You may call it superstition, if you like, but I have always felt him something sacred. I have felt as if he were a mysterious
influence in my own life, shaping it for the highest
things. And at the same time it's as if he appealed to me, always, from his grave, for protec-
tion. Since I was old enough to realize that I had
lost him, I have never been recreant to his trust
in me."

"Yes, I should feel just so about my own father," Hope granted.

Langbrith put aside the comparison of his father with hers by something in his tone rather than in his words. "Yes," he assented, though he refused her sympathy on those terms. "But it isn't the same thing. My father is dead; and, while he lived, he was not a man who could make himself understood; I can't explain; in all the letters he left, and his memorandum-books, it was implied. I thought my mother felt the same, and that was why she was so silent about him; and I thought that Dr. Anther— But if all the time they were conspiring to betray him—if they were thinking of themselves and each other, when they, of all people in the world, should have been the truest to him—"

"Oh, oh, what *talk!*" Hope broke in. "Why, James Langbrith, I should think you were insane."

"I am! I am!" he choked out. "This thing is turning my brain. I try to realize it, and then when I realize it I feel that I must go mad. Oh, you don't understand; you can't! you can't! I feel so covered with shame for my mother."

As they talked, they walked swiftly. Now and then the moon struck between the trees, upon them, but in the prevailing shadow they had the seclusion in which they willingly hid themselves, till they reached the top of the ridge that overlooked the house and below that the town. Its varied murmurs came up to them there, with the sound of the mills vibrating through all.

"I suppose," he said, bitterly, "that they all think

I am a fool to care for him, though he made their prosperity, and did more for them than they did for themselves all together."

"Now, you sha'n't be morbid, if I can help it," she broke out upon him. "I don't believe any such thing, and I don't believe you will, when you come to think. Do you want me to talk up to you the way I used to at school, or to pretend I'm afraid of you, and flatter you and make you think you've been abused?"

"How do you mean?"

"Of course, we haven't been together so much since you went to Harvard; but since—since this afternoon, I've been feeling the old way, as if we were children again, and we should always speak right out anything we thought. There wouldn't be any use or sense in it if we couldn't."

"Why, of course."

"And you believe that I care for you more than for any one else in the world?"

"That's how I care for you."

"And that I wouldn't say anything I didn't believe was for your good?"

"I can't think of you apart from myself."

"Well, then, listen: you know very well that everybody honors you for wanting to commemorate your father. They don't know anything about him, but they think you do; so that's settled, and we won't have any misanthropy for the people down there. Now you're sure I may say what else I think to you?"

"Anybody may say what they think to me."

"Oh, if you want to be boyish!"

"Go on, Hope!" he said, humbly. "I beg your pardon."

"There's no pardon to be begged or granted. I just want you to see this in the right light, and I've got, first of all, to know what you said to your mother."

"I don't remember the words. But I let her know how I felt," he gloomily answered.

"And to Dr. Anther?"

"Nothing! I wouldn't speak to him. But I let him know that he was ordered out of the house."

"You did! Well, I've half a mind never to speak to you again. And did he go?"

"No. *I* went," Langbrith said, with sullenness, somewhat crestfallen. "I told my mother I was going to Boston with Falk, to-night. Did you expect me to stay and see them married?"

"Where is Mr. Falk?" Hope asked, as if to gain time before answering his question.

"I couldn't find him. He was walking somewhere with Susie Johns."

"And why didn't you go to Boston without him?"

He looked into her face in a daze that did not at once yield to her intention.

"Without coming to see you?"

"Oh, you stayed for that, and now it's too late to go."

"It's too late."

"And so you're going back to your mother?"

"I'm going back to the hotel for the night; then—"

"No, James," she said, gently, dropping her mockery in the seriousness which was in the veiled depth of her nature; "you mustn't do that. I want you to do what I say. Will you?"

"I will listen to what you say."

"No, that isn't enough. I want you to go back to your mother, and say, 'I was all wrong; I know I am wrong because I know you couldn't be.' Will you say that?"

"No, never! I wouldn't say it if you made it a condition of my ever seeing you again."

"Do you think I would do that, or do anything that would make me a tyrant over you? I am not so foolish, no matter how wicked I am. I wouldn't give you up if you chose to stay in the wrong; but I know some day you will want to put yourself in the right, and I don't want it to be too late. Now will you do this? Go and say, 'Mother, I can't withdraw what I said, but I know you believe that you are doing right, and I will stay here and be at your marriage.' Will you?"

"Why do you wish me to do that?" he struggled against the sense that he was giving way to her.

"Because I hate you, and want to do you all the harm I can."

He understood. "Well, I can't tell her what you say; but if she wishes to marry that man, she mustn't seem to do it against my will."

"And you'll promise her to be at the ceremony?"

"No, I won't do that, Hope; I won't do it even for you. How could I, without seeming to condone it, to approve of it, when my whole nature revolts against it? Would you want me to act such a lie as that?"

"You know that it would seem a quarrel with your mother if you didn't."

"Well, there *is* a quarrel. She's no mother of mine if she marries that man."

"But you said yourself that she mustn't seem to do it against your will."

"No matter for that. They can wait till I go away; till I put the ocean between me and the loathsome thought of it. I've promised enough. I can't do more than I've said I would; no, not even for your asking, Hope!"

"Do you think I ask it for myself?"

"No."

"For Dr. Anther, or your mother, even?"

"No."

"For whom, then?"

"For me, I suppose. But you ask too much for me."

She could not mistake his sullen finality. She sighed deeply, but not desperately. "Well, then, tell her that you won't oppose the marriage. And if you are going to do that, you can't do it too soon"; and she began to find her way trippingly down the slope that led from the hill-top into the garden behind her house. Langbrith followed more heavily and more slowly, and less securely of the way, and she had to wait for him beside the gate, on which the moon was trying feebly to paint the hour. He felt that he had no right to her embrace; but, when she turned there and put her arms round his neck and kissed him, his sore heart melted within him.

He wanted to say that he would do what she asked, but somehow he could not, and they parted without further words, except a whispered "Good-night!" from him, and a "Good-night, dearest! Be good!" from her.

A light showed in the roof-chamber which he knew, and as he turned the corner of the house, towards the street, a low moaning, the precursor of nightmare within, stole out of the lifted sash on the moonlit air. He thought of the burden and afflic-tion her father was; but he did not think, he was too young for that, what a burden and affliction her husband might be; and, doubtless, Hope herself, strengthened for one trial by the other, did not feel either beyond her woman's force, or both more than her woman's share.

Langbrith pulled himself more and more slowly homeward. The outer door was open, as he had left it, and he passed in and stood a moment at the door of the parlor. The moonlight without showed him his mother sitting in the room, as if she had not stirred from the chair into which he had seen her sink when he madly broke from her entreaty.

"Mother," he said, stonily, for all the pathos of the sight, "I know you think you're right, and, if you're going to be married soon, I will stay and be present."

At first she did not answer; but, after he had be-gun to imagine she had not heard him, she said, "I am not going to be married."

Langbrith waited, in his turn, before he said, "I don't understand; but I suppose you know what you mean, at any rate." And he now felt himself speaking as much to Hope as to his mother; "I've done what I could."

"Oh, yes," she answered, with bitter rejection of the immediate purport of his words, "you've done what you could."

XXVIII

Mrs. Langbrith did not wait for Anther to come to her for the withdrawal of her promise: she could not take the chance of another meeting between him and her son. She sent him word the next morning, as soon as Norah was up. She had not slept, consciously waiting to send it, and he had not slept, unconsciously waiting to receive it.

He read her note without surprise; he read it almost, he felt, with a sort of expectation. "I must take back my word. I cannot keep it. You know why. We ought to have known I could not." Within, the note was neither addressed nor signed. He read it passively, and folded it up and put it into his pocket-book. That day, as he made his visits, he thought recurrently of those weak forms of animal life which gather their strength for a sudden spurt, and then, when it is spent, rest helpless till their forces are renewed. He had taken her in a moment when her will had accumulated strength enough for action; but the impulse had exhausted itself, and now she could not act. At first, he said to himself that he must wait for another rise of her slight powers, and then help her to prevail with herself. But, at last, he said that this would be taking advantage of her weakness—making himself her tyrant, her oppressor.

She was not less but more dear to him because of her feeble will. He had always pitied her for her subjection to the brute force of others—of her husband and of her son; and the love that had begun in pity continued increasingly in pity. She was never so dear to him as now when she had failed him. He could not decide that she had failed him more finally than before, but she had failed him more signally. He promised himself that he would not try to see her again, as long as her son was with her; and, in fact, it was not till the morning after Langbrith had been gone a week that he stopped his horse at her gate, and found his way up the box-bordered path to her door.

She had seen him coming, and met him at the threshold with a dismay and entreaty that went to his heart.

"How do you do, Amelia?" he asked. And she answered:

"Have you come to say that you despise me?"

"Do you know me so little as not to know that I care for you more than ever I did?" he protested. But at the kind of fluttering in her, he said: "I hadn't come to speak of that; I never will be the first to speak of that again. I know that James has left you. I wouldn't come till I knew that, though I wanted to assure myself with my own eyes that you were well."

She looked at him in gratitude that included the larger with the lesser favor, and answered, evasively, "He went last Saturday; he sailed this morning."

"Yes."

"Hope Hawberk and he are engaged. She's been to see me. But he told me before he left. She's a good girl."

Anther said, as if in reply, "James is a good man." Those unfalling tears came into Mrs. Langbrith's eyes.

"*I* know who is a good man," she said.

"When are they to be married?" Anther asked, ignoring her worship.

"They don't know, exactly; not for a year, at least. He says he wants to be sure that he is doing something over there, first. Do you understand what it is that he wants to do?"

"Not very well; but I have heard of other young men studying to write for the theatre, there. The French are supposed to do those things best."

"Yes," Mrs. Langbrith vaguely assented. "He's got that Mr. Falk to go with him."

"That's good. He seems to be a good fellow."

"I think he will be a good influence," Mrs. Langbrith suggested.

"Oh yes, but James could be trusted to himself."

"Yes. Dr. Anther," she broke off, "do you think Mr. Hawberk is going to get well?"

He looked quickly at her.

"Why?"

"Hope thinks he is. She says he is trying harder than he ever did before; he's paying more attention to what you want him to do. She says that the days when you want him to take the medicine instead of the laudanum he does not take the laudanum at all."

"I haven't seen him for more than a week. His gain depends upon how long he has been keeping faith with me."

"I guess it's more than five or six days, now. Dr. Anther, if Mr. Hawberk should get over it, would he begin to tell the truth, or would he go on talking the same as he does now?"

"Talking about what?"

"Oh, everything. You said that his opium-eating prevented his telling the truth."

"The truth?"

"Well," she said, desperately, "about Mr. Langbrith. If he got well, would he say what Mr. Langbrith really was?"

Anther rose, and walked across the room and back; and he did not sit down again. "He would be apt to say what he really was."

She drew a long breath. "I don't know as I should like that," she said, piteously, and her voice trembled. "It would get to James, and — and — I don't know as I want he should ever know, now."

They looked at each other, he searchingly, she beseechingly. He wondered, "What is she asking me?" and a pit, on the edge of which she seemed to tremble, opened to his conjecture. His gaze hardened, and hers sank under it. "I've nothing to do with that," he said to her falling face. "My business is to cure Hawberk, if I can, at any risk, and with any consequence."

She returned wildly, as if in terror of something she had barely escaped. "Yes, yes! You must!

And, oh, I hope you can do it! I can't help what he says about Mr. Langbrith; I don't care who knows the truth. Only cure him! Why do you look at me so, Dr. Anther, as if you blamed me? Well, I *am* to blame. I did—"

"Hush, Amelia! I don't blame you. I understand you. Don't think I blame you, or hold you responsible for anything." Whatever it was that had passed from one consciousness to the other was confessed and pardoned, and he took her hand in saying, while her tears rose without falling, "I have been thinking the whole matter over very anxiously since I saw you last, and I have asked myself, now that we can never be anything more to each other than we are, what would be the use of James's ever knowing the sort of man his father was. I have had my impulses to revenge myself on him, to punish him for what I have considered his insolence to you as well as me; but I have fully realized that his wrong came from the illusion in which he lived, and that we could not destroy this illusion now to any good purpose, and I have no longer any wish to hurt him. Let it go. But as a physician," he added, "there can be no doubt of my duty. Hawberk might live on indefinitely, as an opium-eater, and, again, he might die suddenly. It's my business to keep him alive and get him well."

"Oh yes, I know that. It was because," she entreated, "I thought it would be so dreadful if I thought it that I thought it."

He looked at her with a sad intelligence where she stood wavering. "I understand," he said, and he took her hand, hesitating. Then he dropped it, saying "Good-bye," and left the house.

He drove away hardly aware of anything outside himself, till he was aware of coming to the rectory. Then he realized that he was going to see Hawberk. He was beset by a sudden longing to speak with Enderby, and was staying himself against it in a sense of its meanness and unfairness, when Mrs. Enderby's voice called to him from the yard, where she was gathering some flowers from the blossomed shrubbery. He perceived he had stopped at the gate.

"Won't you come in and see Dr. Enderby?" she called.

"No, no, I thank you," he returned. "I hope he's well."

"Oh, quite well," she answered, looking at the sprays in her hand. "I was just getting some flowers to send to Hope," she said, as she came to the gate. "Aren't these roses magnificent?" She touched their cheeks with the hand from which she dangled her garden-shears. "They're fit for any fiancée, even such a little dear as Hope. You've heard, of course?"

"Yes."

"It's too delightful! There's something very romantic, don't you think? in his remaining constant to her after all the nice girls he must have seen in Boston and Cambridge and Brookline, as a student. But she's wonderful! Yes, she is. And so happy! Have you seen her since?"

"No, but I suppose I shall see her now. May I take her your flowers?"

"Oh, will you? Thank you, so much." She came out and put the flowers on the seat, where he made room for them beside him. "It's rather hard," she ran on, "her being left behind here and he gone out to Paris. But her father's being so much better is a great compensation. You must feel doubly anxious to cure him now. Of course, they never could think of marrying and going away from him while he's in this state. And you really have hopes of him?"

Anther could not smile, even in his amusement with the comely, kindly woman beaming up at him with her hand above her eyes. "I'm doing my best," he said, gravely.

"And you will succeed."

"These cases are difficult; but I have my hopes."

"And you shall have my prayers—*our* prayers!" she said, fervently. "You won't come in and see Dr. Enderby?"

"Not this morning. I have too much to say to him."

"Yes," she assented, dropping her eyes; and he knew that she knew what was in his thoughts.

XXIX

ANTHER did not find Hawberk. Hope said, from the steps of the house which the doctor could drive so near, that he had gone down into the village; she believed that he meant to call at the doctor's office before he came back. She was always a cheerful presence, but now joy seemed to radiate from her like a rapturous effulgence. Anther felt it, and he felt that if she knew of his own reason for sadness, she had the same right to ignore it in her own nerves as she had to ignore her father's misery. She looked as if lifted tiptoe by her happiness, and her voice danced with her dancing eyes.

"Why haven't you been to congratulate me, Dr. Anther?" she challenged him, archly. "I do believe you don't care!"

"Oh, I guess not," he retorted, feeling his load raised in part by the mere ecstasy of her spirits. "I hadn't been officially notified."

"Well, you are now. I was waiting to come and tell you, when I was sure I could tell you how much better father was. He hasn't disobeyed for nearly a week—ever since James left. It all seems to come together. It's made me so wretched."

"Well, you don't look it," he answered. But he did not smile at her mocking, and she recollected herself. She looked at him in wistful sympathy, but the years between them were so many, and Dr. Anther was such a really dignified person, that she could not venture to speak her sympathy, uninvited. He did not invite her. He felt himself blush at the pity of the joyous young creature who was imagining his case from her own, in an equality of passion. It embarrassed him in his consciousness of the difference. She grew a little embarrassed, herself, and he knew that he was wounding her. "Did you say your father had gone to see me?" he asked, gathering up his reins, while Hope stepped back from the wheels.

"Dr. Anther!" The hoarse croak of her grandmother intercepted her answer, and the doctor saw the stooping figure and fierce face of the old woman in the open doorway; "I want you should tell this crazy girl there can't come any good from that Langbrith tribe. I know 'em root and branch, and I don't know any good of 'em. If ever Lorenzo Hawberk gets to be a man again, instead of a laudanum toper, I can tell him a thing or two about the Langbriths that 'll lock their wheels for 'em."

Hope turned and ran back to her grandmother, whom she gently pushed in-doors. "Now, grandmother, I guess Dr. Anther knows as much about the Langbriths as you do," she said, and she turned her laughing face over her shoulder to show him that she was not taking her grandmother seriously. "Good-bye, Dr. Anther," she shouted, and, suddenly remembering the flowers, he called out to her:

"Oh, hold on a moment, Hope; here's something Mrs. Enderby sent you by me."

"Well, I never did!" she rippled down to him with a laugh that denied any sadness in the world. "What *would* you have said if you'd forgotten altogether? Oh, how gorgeous!"

She fluttered up the steps again with her face buried in the flowers, and then she called back, "Oh, *I* forgot, this time. Thank you, Dr. Anther," she sweetly chanted, and the doctor drove away.

He felt it an escape not to find Hawberk waiting for him. He found both the bottles, one for laudanum and one for medicine, which Hawberk had left, and a scribbled note from him: "Will call for these later. Guess we're getting the upper hand of that green fellow a little. I couldn't get him to come with me, anyway."

Dr. Anther was taking his meals at the hotel when he could think of them or time them aright, and his hired man was in a sort of loose, general charge of his place, pending the installation of some specific house-keeper, of whom the doctor had as yet no distinct prevision. When the hired man was not about, the door was free to any one who would open it, and patients came in and waited for the doctor, or wrote their calls on his slate and went away.

He now examined his slate, and found no call so pressing but that he felt justified in sitting down and giving Hawberk a chance to return before he started on his rounds. He was perplexed by a situation which would once have been joy and triumph to him, mixed with the hope whose fierceness he now recognized with abhorrence. What had worn the high look of righteous retribution and been the promise of happiness was now more like a menace of the peace which alone remained his desire, as far as he had any desire. He had been beaten in the struggle. The dead hand had been too strong for him. If he could still prevail, through Hawberk's restoration to truth in his restoration to health, he would prevail in vain, for he would prevail too late. Nothing but his duty remained, a duty that was barren of personal reward, and that if done successfully, as regarded Hawberk, must be done at the risk of fruitless suffering for others. It was with a sense of reluctance close upon disgust that he pulled himself together, at the sound of shuffling steps, which he did not doubt were Hawberk's, loosely dragging themselves up his walk.

John Langbrith came in, and lounged weakly into the easy-chair with a cursory nod to the doctor. "I want to see," he said, without further greeting, "if you can do something for this dyspepsia of mine." They had not parted friends, or even courteous acquaintances, at their last meeting; but, as John Langbrith ignored that, Anther ignored it, too, in the superior interest of their relation as patient and physician.

"Is it worse?" he asked.

"If it wasn't worse, I shouldn't have come. I can stand a good deal without squealing, but I can't stand everything!" Langbrith began nervously swinging the leg he had crossed upon the other, and

275

looked about for something to chew. In default of anything else, he tore a piece from the splint bottom of his chair and chewed upon that, as he laconically, almost sardonically, rehearsed his symptoms. Anther listened without prompting questions, and at the end John Langbrith said, "I presume you'll come out with the old thing: overwork."

Anther rubbed his hand all over his face, after his fashion. "That's usually the trouble with nervous dyspeptics, when it isn't overeating or overdrinking. Couldn't you get a little time off and go somewhere for a change, as well as rest?"

"I guess I've got to. What can you give me to take, while I'm putting things in shape to leave?"

"I'll do something to tide you along; but you understand that it's merely temporary." Anther turned in his chair to write a prescription, pausing and thinking over it, while John Langbrith continued talking to his back.

"If you could get off on a good long sea-voyage, it would be the best thing—two or three weeks."

"I could get off on that as well as anything else. The devil of it is to get off on anything at all. There ain't a soul to leave the business with, the way I want to. If that fool of a boy was worth the powder to blow him, I should be all right. What's he going to do over there, anyway? You make it out?"

"He's going to learn to write plays, as I understand."

"Write plays!" John Langbrith grunted. "Who wants his plays?"

"That remains to be seen."

"Well, I'm not goin' to stan' it. They'll find that out, both of 'em. If his mother hadn't babied him up so, and kept him in cotton all his life, I could have worked him into the business before this, and now I could leave it in his hands."

"You say you don't sleep very well?"

"Sleep! How can a man sleep with a stomach like mine? But I shouldn't care for the not sleeping. Never did want much sleep. The devil of it is, I don't *wake* well. Sometimes I'm in such misery I don't hardly know where I am. Why can't you give me some of Hawberk's dose?"

"I can if you want to come to Hawberk's condition."

"I suppose you could cure me if you have him?"

"I don't boast of having cured him, yet."

"I thought you did, the last time." Langbrith chuckled with a dry pleasure, while he seemed indifferent as to the doctor's sharing in the recollection. "If you could get him on to his legs again, I might leave him in charge of the mills. Maybe he wouldn't want to blow on Royal, then!"

Anther still sat stooped over his desk, and gave no heed to Langbrith's continued pleasantry. He wheeled abruptly in his chair, and held a prescription towards his patient. "There, that's the best I can do for you now; but get away as soon as you can."

Langbrith folded up the prescription, and put it into his pocket-book, but he did not rise at once. "I guess I shall have to, unless this does the business for me. I don't know why I'm so anxious about the damned mills, anyway. Royal always treated me like a nigger—he did everybody he could get under his thumb, and this boy seems to think I'm part of the property. It wasn't for either of them I couldn't meet you on your proposition the other day."

"Oh, that's all right!" the doctor said.

"I shouldn't care if Hawberk came out with the true story some day. But I don't want to go outside of my job, if I don't have to. That's all there is to it. I've got enough to do, running the business, without looking for trouble with Royal's ghost!"

Anther had nothing to offer on this point, and in the country fashion, in such cases, he said nothing at all. And he did not respond in any wise to the long-drawn, groaned-out "We-e-ell!" with which John Langbrith got himself away, as a form of leave-taking. He had been gone some time when Hawberk came in, with a step so much firmer and quicker than Anther had known it for a long time that he could not have known it as his.

"Well, Doct' Anther," he said, briskly, "have you got my bottles ready for me?"

"I've not got your prescriptions ready; I happen to be out of the drugs," Anther said, with a returning sense of meaning in the duty which had lately seemed so purposeless, and a rise of liking for Hawberk in the place of his reluctance and disgust. He felt the charm of the man, which he had never quite ceased to feel, though it had been dulled by long disappointment with him.

"Well, I don't know," Hawberk said, "but I guess we're doing the business for that green fellow at last. I always did know what he was when he seemed to be coming at me by the thousand, like your reflection, you know, when you stand between a couple of glasses. That got to be a great trick of his one while; but he's stopped it now. Why, Doct' Anther," he exclaimed, with a sort of impersonal pleasure in a fact which Anther must enjoy, "I've got so, inside of the last forty-eight hours, that I haven't been afraid to go to sleep. He still keeps hanging round, but he seems to know I'm on to him, and he don't try any of his old jinks with me; just comes and goes to let me know he's around, but don't make any particular trouble. Why, doctor, just to try myself, one day this week—day before yesterday, I guess it was—I got down to sixty drops of laudanum, and it was my laudanum day, too. Don't I show it—in my looks, I mean?"

"Your complexion is clearing up. But go slow, Hawberk, even when you are going in the direction of my instructions. I don't want you to tamper with my patient's case."

Hawberk tasted the humor. "Well, I won't, doctor; I won't," he said, and he laughed in the free way that was natural to him, and that went to Anther's heart.

The doctor turned a little grave, though. "How are the psychological symptoms? Do you see things, generally, as you have been seeing them?"

"I don't know as I do—everything. There's one thing I wanted to speak to you about."

"That house you're going to put up on the hill back of you?"

Hawberk smiled. "I guess that can wait awhile." Then he said, seriously, "You know Hope and Jim Langbrith have fixed it up between them?"

"Yes, Hope told me this morning. I had heard of it before."

"Well, that's all right. He's a good fellow, and I haven't a word against him. I don't know what he's going to do out there in Paris, but I presume *he* does. Anyway, Hope believes in it, and if it never comes to anything, he's got money enough without it." Hawberk's face clouded. "I suppose if everything had gone right, I should have had some money, too. That's the way it looks, off and on. I've had times, of late, very curious times, Doct' Anther, when it don't seem as if the square thing had been done by me. Do your remember the circumstances of my leaving the mills? I ain't clear, myself."

The dawn in Hawberk's mind had broken sooner than Anther expected, though it had come too late for any purpose of his. Now, if he had a wish, it must have been to darken it. When he thought how he would have once exulted in it, he had a kind of sickness of it; but his duty was still before him. He must do his best to cure his patient. As a physician, he could have no other concern; but he could keep himself out of the moral consequences. With these he had nothing, and must have nothing, to do.

"You'll be clear enough if you get well," he said. "All the facts of that matter are something you must work out for yourself. I wish to caution you only on one point. You must be very careful to verify any surmise you may have. I should urge you not to speak of it to any one but me. You can see how, under the present circumstances, it could make great unhappiness for James Langbrith, and through him, for Hope."

"Yes, I see that, doctor. I'm not speaking of it. As I say, it's something that comes and goes." He added, with a laugh, "And it goes full as much as it comes. Well," he rose and took his bottles from Anther's table, "Emmering put these up for me? You ain't afraid we'll get our heads together and make 'em *both* laudanum?"

"I guess I can trust you," the doctor answered, almost absently.

"Well, I guess you're right. Anyway, that green fellow has got the job of watching after me, and he's on the lookout. You've weakened on the laudanum a little this time, as I understand."

"A little."

Anther's absence gained upon him so much that he scarcely noticed Hawberk's going; and he sat long in a hapless muse, in which now and then he smiled in self-derision. If the situation had been contrived by the sardonic spirit of Royal Langbrith himself, it could not have had a more diabolical perfection.

XXX

Life is never the logical and consequent thing we argue from the moral and intellectual premises. There ought always to be evident reason in it; but such reason as it has is often crossed and obscured by perverse events, which, in our brief perspective, give it the aspect of a helpless craze. Obvious effect does not follow obvious cause; there is sometimes no perceptible cause for the effects we see. The law that we find at work in the material world is, apparently, absent from the moral world; not, imaginably, because it is without law, but because the law is of such cosmical vastness in its operation that it is only once or twice sensible to any man's experience. The seasons come and go in orderly course, but the incidents of human life have not the orderly procession of the seasons; so far as the sages or the saints are able convincingly to affirm, they have only the capricious vicissitudes of weather.

Anther had been in charge of Hawberk's case for twenty years; and, though he had always forbidden himself to despair of it, he had long ceased to hope for any final cure. He was used to changes for the better and changes for the worse in Hawberk's habit, and to the psychological consequences when he limited his indulgence and when he lapsed again into his debauch. Under it all, though the man's character was deteriorated or ameliorated, his temperament remained fundamentally the same, and Anther had never ceased to feel his gayety and his goodness, which, as they reappeared in Hope, charmed and deeply touched him. Hawberk's recovery had become personally indifferent to him, so far as it concerned the hopes he had once built upon it; but the girl's joy in it gave poignancy to the fears that had replaced his hopes. In a reasonable forecast of the effect, Hawberk must return in his self-restoration to a full sense of the reality concerning the wrong done him by Langbrith; and in place of the delusion he had promoted in the helpless mendacity of his habit, he must know and speak the truth. There had already been hints of such an eventuality; the hints that sickened Anther in his thought of the time when he would have welcomed them, and that made him tremble for the misery which the truth must bring upon Hope, through her love for the man whose father had so pitilessly wronged her own. Anther had believed that he wanted justice done. This had been his argument with Judge Garley; it had been his suggestion to Dr. Enderby. It ought to avail him in any emergency, but now it did not avail him, and he accused himself of having cared for the truth only in his own interest, as the truth would have promoted it with Mrs. Langbrith against her son.

What did avail him in the course he must pursue was his sense of professional duty; amid all the moral confusion, that was clear. He ought to have no question but of the recovery of his patient, and he tried to fix his mind upon this, and not let it stray to any question of consequences. He did his best to keep his study of the case physiological, and not to concern himself with those psychological aspects which Hawberk himself found more interesting, and which he was fond of turning to the light in his visits to his physician. With his escape from the terrors of his opium nightmares, he found a philosophic pleasure in noting facts from which even the physician was aware of shrinking.

Once, towards the end of summer, when they had been "taking stock," as Hawberk called it, of his symptoms, and he was exulting in the reduction of his laudanum to the equivalent of three grains of morphine a day, he said: "The most curious thing about it is that I seem to be doing a sort of Rip Van Winkle act, and waking out of a dream of twenty years or so. It's a dream that's been going on steadily all the while that those little one-horse nightmares have been cavorting round, with green dwarfs on their backs, and playing the devil generally; and this steady dream has had a good genius in it that I'm beginning to have my doubts about, now that I'm waking up. It seems to me that Royal Langbrith wasn't such a friend of mine as I've been trying to make out. What do you think? Or did I put this up on you once before?"

"Not just in so many words."

"Well, I wasn't certain. Royal Langbrith seems to have a better grip as a good genius when I've been dipping into the laudanum pretty freely than he does when I've kept to the medicine and the tonics. I have my ups and downs about him. But what do *you* think of him in the capacity of a good genius?"

"As I told you before, Hawberk, that's something you've got to work out for yourself."

"And if I've worked it out that he was an infernal scoundrel, and was ready to say so, what are the chances that folks would believe it?"

"The chances would be against you, with your past as an opium-eater."

"They could say it was another of my pipe-dreams?"

"You would have to bring the strongest sort of proof."

"With every one?"

"What makes you think now that you were mistaken about him before?"

"Look here, Doct' Anther, what do *you* think about Royal Langbrith?"

Anther suddenly perceived that he had a duty towards Hawberk not contained in the duty of a physician to his patient: the duty one has to a man whom one knows to have been wronged. "I?" he hesitated. Then he plunged. "*I* think he was an infernal scoundrel!"

Hawberk laughed queerly. "Don't you *know* he was?"

"Yes, I *know* he was." The truth was open between them, and each was astonished at the effect the open truth had on himself.

"What," Hawberk parleyed, with a smile as queer

as his laugh, "should you say we'd ought to do about it?"

"I don't know," Anther candidly avowed. "Once I should have known."

"So should I." And now Hawberk roared with pleasure. "But I guess that devil has *got* us now. I've seen the time when I wanted to go into the cemetery and dig him up and burn him, but I don't know as I do now. What do you say, Doct' Anther? Let by-gones be by-gones, as the fellow said about his old debts when he started in to make some new ones? Still, it does gravel me when I think of that tablet in the front of the library. I was looking at it as I came along down. Kind of pathetic, too, when you think of Jim. How did they ever keep him in the dark about his father?"

"It happened naturally enough. It rested with his mother; and, when the time came for him to know the facts, the time for her to tell them was past."

"I see. A good deal as it is with me now. You might almost say that devil had planned it out to have his boy make it up with my girl, so as to stop my mouth for good and all. First off, after I lost my wife, I used to think I should like to make him suffer for the lies he threatened me with. I wanted to kill him. Well, what's the use? Somehow, I don't feel that way now. I don't want to revenge myself, and I don't believe she'd want me to revenge her. Curious!" Hawberk reflected, with a pause, in view of the interesting predicament. After a while he said, "How that devil must have chuckled when he saw me up there, with the other leading citizens that day, dedicating that tablet to his memory! But, Doct' Anther, there's something I can't get through me. I can understand why *I* should be there. *I* was game for anything, when I was filled up with laudanum; but I don't see how you came to be celebrating the life, death, and Christian sufferings of Royal Langbrith. Never did *you* any harm, did he?"

"Not while he lived," Anther said.

"Kind of fetched you a back-hander from the grave? Well, I don't want to ask you what it was, but I should like to ask how you came, knowing all you did about him, to let Judge Garley and Dr. Enderby in for their share in the proceedings. They any notion of the peculiar virtues of the deceased?"

A painful flush overspread Anther's face. "I felt it my duty to tell Judge Garley as soon as I found that the scheme had taken shape in James's mind, and he held the legal view of it. He was duly warned, and I have nothing to blame myself with there. I don't feel so easy about Dr. Enderby. I am afraid I let a personal motive influence me in withholding the truth from him until it was practically too late for him to withdraw. I can't decide how much he wished to spare me in arriving at the conclusion he did. He agreed substantially with Garley that no good could come of exposing Langbrith at this late day, and much harm might come. Besides, James was to be considered."

"Ah!" Hawberk said. "That's where *I* come in. What about James? Hadn't he ought to know about it? Hadn't I ought to have it out with him before he marries a daughter of mine?"

"Dr. Enderby thought that no one should tell him now; that no one could, without interfering with the order of Providence, without forcing God's purposes, as he put it. The truth could come out fully only when it could come out naturally, necessarily, inevitably."

Hawberk fetched a long, deep sigh of relief. "Well, that lets me out. I was feeling my way in that direction, I guess. I guess Doct' Enderby is right. Any rate, I'm going to let the thing rest for the present. I'm satisfied with what I've got. It wouldn't help me any, and it wouldn't help Hope, if the whole thing was out. Let the damned thing be, *I* say, and that's what I understand Doct' Enderby says: maybe not just in the same words. I don't know as I should exactly want Hope to marry Jim Langbrith, without he had been told something about it — say enough to understand that there wa'n't any flies on me when I was put out. That's only fair to Hope; I don't care for myself. But if there's an order of Providence, I'm willing to wait for the procession. Yes, I'm willing to wait and see if there *is* any procession. If there ain't, it'll be time enough to start one. Well, Doct' Anther," Hawberk said, putting out his hand to the doctor as he rose, "I don't want to holler before I'm out of the woods, but as far as I'm a judge, you've saved me, body and soul. I don't know how you feel, but I should be glad to swap my feelings for yours, whatever they are. Yes," and Hawberk broke down with his laugh from the height of sentiment he had reached; "I don't know but I'd be willing to swap Royal Langbrith's feelings for yours, this minute."

Anther could not refuse to join in his laugh, but he felt it right to put in a word of caution. "We mustn't brag about your case. But I'll say that I've hopes of you that I never had before. It now rests with you, mainly. If we pull through together, I'll be glad to swap feelings with you. We won't say anything about Langbrith; he mightn't be willing to trade."

"Not without some boot, you may bet," Hawberk shouted, with supreme joy in the joke, as he went out of the doctor's door, where the doctor stood looking after him, not unhappy for himself, as he ought logically to have been in contrasting his hopeless life with the life that was beginning anew so hopefully for Hawberk, and with something of the peace that passes understanding in his heart.

XXXI

JOHN LANGBRITH continued to talk of going away.
Upon the inspiration of meeting an old acquaintance whom he asked where he had been keeping
himself of late, and who answered that he had been
in Japan, John Langbrith began to think of going
round the world, as a little experimental journey,
since a man could go to Japan and back without
being noticed. He asked Anther what he thought
of circumnavigating the globe as a remedy for nervous dyspepsia, and the doctor told him he did not
think it would be bad. Then John Langbrith said
he had half a notion to go out to Paris, and see
James; there had never been much affection between
them, but John Langbrith considered that James
could get him a comfortable boarding-place, where
he could stay while he was picking out some German
spring to go to more permanently. He asked Anther if he did not think some of those German springs
would be good for him. Again Anther said that he
did not think it would be bad; and this suggested
giving Saratoga a trial. John Langbrith could go to
Saratoga for a week before the season ended, and he
shaped his business so that he could put it in the
hands of a young subordinate, with instructions to
reach him by telegraph if needed, for he could re-

turn at a second's notice; and he actually went. At Saratoga he drank impartially of all the waters, at all hours of the day, without regard to diet, and came home worse, if anything, than he went, but somehow with a sense of renewed energy.

He took hold with so much force that, before the snow flew, he had, as he phrased it to Anther, got round to a little back of where he started. Then the doctor indulged a sentiment of something like poetic justice, in suggesting a means of relief for John Langbrith from one side of his work, and of benefit for another patient.

"Why don't you split up your responsibility?" he asked. "Shoulder the business half yourself, and let Hawberk look after the manufacturing. He needs something to help keep him out of mischief, and he is able now to take hold of the paper-making and run it as well as ever he did. He hasn't forgotten how to use his own inventions, I guess."

John Langbrith's jaundiced eyes emitted a yellow light of appreciative relish. "Lord! Make Royal turn in his grave—what there's left of him *to* turn! Do you mean to say you could put any dependence on Hawberk?"

"Why not? It would be merely a mechanical exercise of his faculties, and it would occupy him and keep his mind off the opium."

"Lord!" John Langbrith said again; and after a moment's muse he said, "Send him round," and so took himself away with a galvanic activity that supported him in his automatic progress towards the mills.

Hawberk had much the same sardonic pleasure as Langbrith had shown at the notion of his being reinstated in his old charge; but it was sweetened to something better by the virtues of temperament in him. "Now, Hope," he bade his daughter, after the first day's experiment had justified the confidence with which he entered on his work, "you write to James about this. He'll like to hear about it, and he'll like to hear about it from *you*. And you tell him it was Doct' Anther's idea. He'd ought to like that, too, and the doctor'd ought to have the credit of it, anyway. If I should make a slump, later on, I'll take the credit of *that*. But I guess there ain't going to *be* any slump."

The few spectators of Hawberk's experiment who could witness it with a fully comprehensive intelligence of the case regarded it according to their respective natures. To the community at large, it had the interest of something miraculous—something between rising from the dead and returning cured from an inebriate asylum. If anything could have rendered Hawberk a more dramatically notable member of society than he had been as an opium eater of twenty-five years' standing, it was his novel quality of reformed opium-eater. This gave him a claim upon the wonder of every stranger who came to Saxmills, and it conferred the right on every citizen to point him out to the sojourner in his going and coming. The fascination of the fact extended itself to Hope, when she happened to be seen, and to the house where the Hawberks lived.

The general belief was that the thing would not

last; and this was the particular belief of Judge
Garley, who owned his scepticism to Dr. Anther,
with some tendency to an amiable criticism of Anther's share in the affair. He had seen so little of
reform, in his acquaintance with the law, he said,
that he was shy of it wherever he saw it. But he
was willing to give it time; it never took much time.
Perhaps, though, he suggested, this was a case not
so much under the law as under the gospel. If that
was so, he would like to know if the doctor really
believed in the supernatural.

"No," Anther said, "only in the natural." And
this was, substantially, the answer which he opposed
to Mrs. Enderby's secret wistfulness regarding a
fact which she beheld as with clasped hands, uncertain how, as a church-woman, she ought to feel towards miracles post-dating those of Scripture. She
would have liked to feel the hand of God in the tardy
and partial retribution of a man cruelly wronged;
and it is doubtful if she thought the rector quite
level with his spiritual opportunities in his preference of Dr. Anther's theory, that the unexpected
was one of the things always to be looked for in the
practice of medicine. What measurably consoled
her was the tender seriousness of her husband in
the whole matter—the brotherly affection which he
showed Hawberk in the relation which he was able
to form with him, as a man doing a man's part in
the world's work after long uselessness, and the
delicacy with which he forbore to recognize that
there was anything novel in this performance of
duty by Hawberk. She was peculiarly touched

when he proposed that they should have Hope and her father to supper, and she promised that she should be forever ashamed that she had let her husband think of it first.

Mrs. Enderby atoned, as far as she could, by asking Mrs. Langbrith and Dr. Anther, but neither of them could come, and she wasn't sorry that they had the Hawberks alone; with retrospective prevision she perceived that anything else would have been overdoing it. She found Hawberk very entertaining. He talked frankly of getting back to his old work in the mill, and he tried to make her understand an invention he had hopes of perfecting for the "Dandy Roll," as he called it, so that the water-marking of paper could be done at an immense saving of time and money. He explained to her that the words, or designs, to be water-marked had now to be put in by hand with bits of fine wire, and sewed on a cylinder with fine metallic thread; but he was trying to make a Dandy Roll on which the design could be changed as easily as if it were a section of type in a printer's form. It was very luminous while he talked, but it all faded away afterwards, and left in Mrs. Enderby's intelligence only the words "Dandy Roll," which had a queer fascination, together with a sense of Hawberk's dignity and enthusiasm about it.

Hope was gay, as always; but it seemed to Mrs. Enderby that she was not so gay as she had sometimes seen her, when she had far less reason to be so. There was a shadow of anxiety in her beauty which Mrs. Enderby wondered never to have found there before, and a sound of anxiety in her lovely tones unheard before. She thought she could see the girl closely following all her father did and said; but perhaps it was only the effect in her of hopes not cherished till now, naturally betraying themselves in anxieties. As a matter of fact, Hope had no reason to feel anything but joy in her father's restoration to his old usefulness. There was no poison of a gratified vengeance in her heart, for it was agreed almost tacitly between Hawberk and Anther that no good could come of her knowing, for the present at least, the outrage of the past. "Time enough," her father had gone so far as to say, "for Hope to be brought into all that when we see that it's got to come out generally. I don't know as I should feel just right about letting her keep on with Jim, if she was one to blame a man for what she has to suffer instead of for what he has done. Any rate, till we see our way to telling Jim, I guess we'd better keep dark with Hope, heigh?"

Whatever might have been the full mind of Dr. Anther, he assented to Hawberk's decision, though he had to hold to it against counter reasoning that searched his deeper nature or his complexer conscience. It was not finally strange to him that this reasoning should have come from one whose peace was more intimately involved than that of any one but Hope herself. Anther must long ago, if it had not been for his tenderness of her, have owned that Mrs. Langbrith had shown a moral cowardice concerning her son, which was hardly less than a culpable weakness; but he defended her to himself, because he perceived that weakness could never be culpable. He might as well blame any of the feeble creatures which she made him think of for not being strong, and he was not ready with praise for the unexpected force which she showed, where he took her weakness for granted. He merely reflected that he had not taken into account the pity of women for women, when one of them has been able to put herself perfectly in another's place, and to ignore in behalf of their sex's helplessness the other claims of nature. A sense of this awed him at Mrs. Langbrith's refusal to acquiesce in Hawberk's notion of what was best to be done in regard to Hope. At first, she had seemed to acquiesce in it, as something that superiorly concerned the father and the daughter. Then one day, suddenly, she went to Anther, and, not finding him, she left a message of peremptory entreaty for him; and they found themselves together, in the early falling twilight of an autumn day, in the dim parlor where their middle-aged drama had already seemed to play itself out.

"I can't let this go on, Dr. Anther," she said, traversing any pretense of greeting between them when he appeared. "Mr. Hawberk is making a mistake. Hope ought to *know*. She ought to be *told*. James is his father's *son*. He may be *like* him. He may make his wife suffer what his father made me suffer. How do we know what he is doing there in Paris, now?"

She was a woman of few words, and in these few

she had compacted her suspicions, her reasons, her conclusions; and, though she pressed them upon Anther with hysterical nervousness, he had to respect the sense there was in them, as well as the anguish there was behind them.

He could only parley, for a beginning. "He is *your* son, too, Amelia."

"And what if he is?" she retorted. "What is *me* in him will be crushed out by what is *him* in him," and Anther saw that she had thought it better than she could speak it, though but for her erring grammar it was spoken well enough.

He said, "I should not fear for her in her marriage with James. She is a stronger character than he."

"That was what I said when I began to think of it. But the weakest man can make the strongest woman suffer things worse than death; and I don't care whether there would be any suffering or not. There would be wrong. She has a right to know. Her father has no right to keep her from knowing. Why, it's wicked! What will she think, what will he say if she doesn't find it out till afterwards?"

"He can say that he didn't know himself. She will not blame him, at any rate."

"That isn't enough. She has got to have the right to say now she will not marry the son of such a man. Will you tell her?"

Anther reflected. "No, Amelia," he said, "I don't think that I will tell her."

"Why?"

"Because I have only the relation of her father's physician to her. If I could have had another relation to her," and Mrs. Langbrith winced at the implication, so that he felt sorry for it, "I might have been justified in telling her. As it is, I don't."

"Well, then," Mrs. Langbrith said, desperately, "*I* will tell her."

"Before you tell *him?*"

The question daunted her; it was necessary, but he realized its cruelty as well as its necessity. She gasped inarticulately; the unfalling tears started into her eyes. She had, as he saw, reached the limit of her small strength. It must be days or weeks, possibly months, before she could gather force for a new effort.

Anther tried to say something consoling to her; he succeeded only in saying something compassionate, which did not avail. "You have taken away my chance," she said, and he would not take from her the slight stay she found in her resentment.

XXXII

ANTHER noted in himself, with curious interest, the accomplished adjustment of the spirit to circumstances that once seemed impossible, and the acceptance of conditions which before had been intolerable. He had gone on to the end of a certain event, strongly willing and meaning something which then he no longer willed or meant. With a sense of acquiescent surprise he found himself at peace with desires and purposes that had long afflicted him with unrest, and it was not they, apparently, that differed, but himself. To the young this will be a mystery, but to those no longer young it will be of the quality of many experiences which, if still mysterious, are not more so than the whole texture of existence.

He had foregone a hope that had seemed essential to his life, but that, once foregone, was like other things outlived—like something of years ago, of his early manhood, almost of his boyhood. He was still baffled and disappointed, but he perceived that he did not care, did not suffer, as he supposed he should care and suffer. It was his compensation that what was ignoble in his regret was gone from it. Neither resentment nor the selfish sense of loss tinged it. Primarily, his regret was hardly for himself; and he perceived that, so far as it concerned another, it was mixed with a sense of escape from anxiety, from fears which the fulfilment of his hopes would have perpetuated. He realized more and more that he had been having to do with weakness, and he realized this not in contempt of weakness, but in the compassion which was the constant lesson of his calling. He blamed Mrs. Langbrith, in her shrinking from collision with her son's will, no more than he would have blamed any timorous creature for seeking to shun a physical ordeal to which it was unequal. He had, at least, learned patience and mercy from his acquaintance with disease; and he had learned to distinguish between what was disease and what was an innate fault which no drugs, either for the soul or body, could medicine.

What surprised him and, when it first suggested itself, shocked him, was a sort of reason, which was not an excuse, for Royal Langbrith in the defect which he realized. Given such a predatory nature as his, was it not in the order of things that there should be another nature formed for his prey? Must not the very helplessness of his victim have been the irresistible lure of his cruelty? We are not masters of those vagaries, good or evil, that fill the mind after its disoccupation by direct purposes; and Anther did not seriously blame himself for their wild play. He broke this up and banished the vagaries sometimes by calling to his help things that he ought to think of, or by confronting them with the woman they wronged and so rendered the more tenderly dear to him.

She was, in fact, never more tenderly dear to him than now, when he had abandoned the hope, almost the wish, of making her his wife. She had been a wife long ago, and yet he began to feel a sort of profanation in the idea of making her a wife. The time came when Anther wondered whether he had ever really felt a passion for her, such as even in middle life a man may feel for a woman, and whether, in that embrace into which they had once been surprised, there was any love other than the affection of a brother and sister, drawn heart to heart in a moment of supreme emotion. At such a time he made entire excuse for James Langbrith, and accounted for him as forgivingly as for her. If her son had instinctively the feeling which had tardily worked itself out in Anther's consciousness, then, surely, it was not the son whom he could blame. One hints at cognitions which refuse anything more positive than intimation, and which can have no proof in the admissions of those who deal conventionally with their own consciences. It was because Anther was not one of these that he was a nature of exceptional type, and because he could accept the logic of his self-knowledge that he was a character of rare strength. He was strong enough not only to forgive the frantic boy who had insulted and outraged him in his pain, but to feel a share in the error which had kept him in ignorance of the truth. It was not the less his right to know this because there had never been the moment to make it known to him. Anther realized that the boy had been deeply injured, and he accepted his own share of the retribution as the just penalty of his share in the error. He saw, too late, that it was his weakness not to have overruled the weakness which he spared the supreme ordeal. He promised himself, somehow, sometime, to make good to James Langbrith the wrong he had suffered.

In this self-promise, after the experience which had stirred his life to its depths, he found a limpid peace from which his dream of passion hung retreatingly aloof, like a cloud broken and drifting away. He had a gayety of heart for which he did not logically account, but in which he felt the power of consoling and supporting the weakness he had once imagined protecting through a husband's rights. When he first saw Mrs. Langbrith after his tacit renunciation, much more real than that explicit renunciation which preceded it, he was aware of an apprehension in her which it was not for words to quiet. By what he forbore, he must make her know that he had ceased to think of her as he had thought, and that she was as safe from the pursuit of what had been his love as from the reproach which he would never join her in making herself.

They talked of Hope and Langbrith, and of the reason there was in believing that it might be safe for the girl to trust her father to himself, if James wished it, before very long. Mrs. Langbrith did not know directly of her son's plans and purposes. Apparently, the communication between them was formal and restricted, and she spoke of what was in her mind rather because of the girl than of him. In an involuntary measurement of her interest with his own, it appeared to Anther that it was he who was the more concerned for James Langbrith; and it was with surprise that he saw she really did not understand him at first when he said, "I wish he could be assured that, when he comes home, there will be no question of its being the same home to him that it has always been."

"I don't know what you mean," she returned.

"I really believe you don't," he said, musingly, with his unselfish gaze on her. "Well," he explained, "that he need not be afraid of my making a difference in it."

"Oh!" she evaded whatever challenge she might have fancied in the words, "he will have a home of his own. Dr. Anther," she continued, "I don't know what you'll think of me, but I don't feel the same towards James that I used to. I can't make it out, exactly, but should you think it was wicked if I had changed so that I did not care for him so much? When I was a child I was that way, if ever they made me do what I didn't want to do, and didn't make me see the reason. I remember it about my mother once, when I was quite little. I had to do what she made me, but after that she wasn't the same to me. It is so with James, now. He is not the same to me. I don't want to punish him for it, but he is not the same. I don't know whether I explain it."

"Yes, I think you do."

"And do you blame me?"

"No, but I think you may change again towards him." She shook her head doubtfully. "You're one of those who need to get back their strength when they have been tried."

His pitying intelligence was very sweet to her. "If I tried to say what I though of *you*—" she began.

"Don't try," he said, simply, and she did not.

She said: "I don't like to think how you have to live there in that way, taking your meals out, and your house so uncomfortable."

"Is it uncomfortable? I don't notice those things very much. I like going to the hotel; it gives variety, and it seems to me I don't get things so cold as I did with Mrs. Burwell."

She gave a house-keeper's sigh of compassion, but she said, from a higher feeling, "I know why you bought it."

"Yes, I told you. But that's all past now."

"Why is it past?" she demanded, almost resentfully. "Do you think I've changed towards you, too, Dr. Anther?"

"No, I don't, Amelia. I believe you're just what you always were towards me."

"Then, if it's all past, as you say, it must be you that have changed."

"No. I am the same, too."

She looked at him with a wistfulness which he knew to be entreaty of him for that strength to give herself to him which she did not feel in her own will.

"If you say so," she tried her courage, "I will do it now—to-morrow—to-day, if you say so. I told you that James took back what he said; that he was willing. At any rate, what is the use? He can never feel right to me after this, no matter what I do. I know him—he can't forgive the hurt to his pride."

"It was a hurt to something better than his pride," Anther said, justly.

"No matter. It's something he can't forgive me and I don't care. You're more than James is, and now he doesn't want me—he won't need me. If you ask me now to marry you, I will."

He believed that he saw in her the little maximum of her force, which perhaps spent itself in the words and would have nothing left for the deed. The deed must be altogether his. In the sweetness that welled up in his soul from the consciousness of perfectly comprehending, not her intention merely, but her nature, he was happier than the fulfilment of his hopes could once have made him.

"Do you say that, Amelia, because you wish it or because you think I do?"

"I want to do everything that you want me to."

"Then I don't want you to do this, my dear. I know you will understand me. I don't believe we ought to get married."

"Because James—?"

"He has nothing to do with it now. Because we can be more to each other if we remain as we are."

She looked bewilderedly at him, but he believed that he saw in her the relief that weakness intimates to one who forbears demand upon it. She had fulfilled her impulse, and spent all her force on it.

305

She was not hurt, either in her vanity or affection. He could see, indeed, that she trusted him too entirely for such an effect.

"Then," she said, in simple abeyance to his judgment, "will you let me do anything for you that I think you need?"

"What is there that I need?" he parried her question. "I am very well as I am. I assure you that I am quite as I wish to be. I don't feel what seems to you discomfort, and after this understanding, that has no misunderstanding in it, I shall feel happier about you than I have ever felt. If I didn't believe you would rather live your life alone, or if I could believe you wanted me to join mine with it for any help I could give, you know I would make you do what you have offered to let me. But I believe the one thing, and I don't believe the other. I know you're wanting to put yourself under my will —to sacrifice yourself to me."

"No!"

"Yes, it is so. If you ever want my help or counsel or friendship, you know it is always here for you, as fully and freely as if I were your husband—perhaps more so. At any rate, I should not exact anything in return, for I need nothing!"

"But if you ever do need anything—me or anything I can do—will you promise—promise—"

"Oh, yes, I will ask you. I promise you that."

Nothing seems final in human experience, and neither of these two who now parted really accepted the conclusion to which they had come as the last word in their affair. It was to be held in that sort

306

of solution in which all human affairs are held, until
that happens which can alone precipitate them.
She went on with the life to which alone she was,
perhaps, equal. She was, at any rate, inveterately
used to its abnegations, if they were abnegations;
and he did the daily duties which were always full
of interest and had the variety which keeps men
from stagnating. He had not falsely pretended
that he liked meeting the new people he met at the
hotel, and he was richer in old companionships than
most men of his age. The new people, it must be
confessed, were oftenest the commercial travellers
whose enterprises brought them to Saxmills. But,
to a man who took other men as unconventionally
as he offered himself, they were less typical and more
personal than they are in common acceptance. The
younger ones might be noisy in manner, and over-
jocular with one another at table and in the hotel
office, where Anther sometimes paused for a moment of digestion after his meals, before driving off
on his calls. But with the old fellow, whose bounds
they did not try to traverse, they were quiet and
gentle. When they had identified him, through the
landlord, they liked to ask him if there was much
sickness around. Now and then, one submitted a
malady of his own to Anther, and took his medicine
with a deferential inquiry whether the doctor
thought smoking hurt a man. Now and then,
there was a young family-man among them, who
was homesick for his wife and babies. The older
family-men liked the quiet of Anther's willing talk,
and put before him their own philosophic conject-

ures and conclusions about life in general. Of their own sort of life they were confessedly tired, but what, at their time of day, could a man do? If they could get hold of a piece of land near a good market, they would be all right. What about abandoned farms in that neighborhood?

Among the transients there happened people who had chanced stopping at Saxmills because they had a fancy for seeing what such a place was like. They were people of independent tastes, from some of the larger cities, and of æsthetic occupations or none, who brought the waft of a larger life and the eagerness of a sympathetic intelligence. There was once an elderly couple from the West, who, after sparely owning that they were originally from this part of the country, developed into pilgrims to the old homestead of one or other of them, which they thought of buying back and fixing up for a summer place, if they could get the children to see it the same way. More than once there was a young couple, still in the flush of immediate marriage, who were breaking their wedding journey to Portland or Montreal or Boston, and were first diffident and then confident of Anther's good-will in his approaches to their acquaintance.

Besides all these, there were regular boarders, as the bank cashier and his wife, somewhat arid financial and social types; and that young and foolish matron who seldom fails, in any village community, to supply food for general reflection, and who, in the idleness of the hotel, where her young husband, a travelling man, has left her, amuses herself by wearing a white yachting-cap and a toothpick about the verandas, and varies her monotonous leisure by buggy-rides with a merchant of the place old enough to behave better.

Anther liked to drop in on Judge Garley of a late afternoon, when he commonly found the jurist reading a novel; he preferred the translations of French novels, which he devoured insatiably, but was as fond, in another way, of scientific tracts, such as he found in the mustard-colored Humboldt series; he liked psychology in any sort and size. With Anther he had always a certain effect of consideration, as one to whom, if not apology, tenderness was due, because of his peculiarities of temperament. The Langbrith incident remained closed between them, and there was no reason for Anther to believe that Garley had any misgivings as to his own attitude in it. Such spare reference to that business as Anther permitted himself was in his talk with Dr. Enderby, whom he fancied of an uneasy mind concerning it, and with whom he had a humane interest in administering the anodyne of his own final peace. It was, in fact, from the rector's reasoning to the conclusion he had reached before that Anther was most helpful to his friend; Enderby himself was never so much satisfied with being in the right as sure that he was right in what he had done. It was one of those experiences, he once owned, that intimate a less perfect adjustment of the moral elements in this life than we may hope for in the life hereafter; as if the earthly materials of conduct were cruder and coarser than the spirit which dealt with them, and which was attuned to finer issues of behavior. Occasionally he asked if Anther knew anything of James Langbrith's immediate purposes, and if he might be expected to return at all soon. He betrayed that he was not at rest with regard to Langbrith's unwittingly making another a sharer in the responsibilities which he must some day assume towards the past.

Mrs. Enderby kept herself as fully instructed as possible from Hope as to the future of the young people, and if she partook of her husband's uneasiness, she did not show it. Perhaps, in that optimistic view of marriage which some of the best women take voluntarily, if not instinctively, she looked forward to that as the panacea of whatever ills life had in store for them. Of course, she allowed, Hope ought somehow to know the truth before she committed herself to the keeping of such a man's son, but this she felt would be somehow divinely rather than humanely accomplished; in reverting to the comfort of a more positive faith from her ancestral Unitarianism, she grew constantly in the grace of a belief in, at least, subjective miracles. That everything would come out right in the end was so clearly a part of the universal justice that she could not have final question of it. When she permitted herself to join in any of the rare and guarded approaches of Anther and her husband to the matter, it was to interpose herself between what the doctor might say and its effect upon the rector. She made herself the interpreter of Anther's acquiescence in the rector's reasoning, so that it should be more of the nature of a robust and positive support. If it would not have taken from Enderby the honor of being first to reach a right conclusion, she might have argued that Anther had himself intimated it to him—when she was less confident of it she sometimes conjectured this. But, for the most part, she was sure that Dr. Enderby had been inspired to it, and that the notion of patience, of waiting on the Supreme Will, of looking for what the older theology called a "leading," was the true ground to take. She was the more to be praised in this because patience was not one of her innate virtues, and it was ordinarily her practice in life to anticipate the signs and tokens for which she was now willing to trust.

Something, in fact a great deal, she held, was to be hoped from Hawberk's return to health and work. There, she argued, was proof that the case had never really lapsed into forgetfulness with the Power that makes for righteousness. It was affecting, it was enough to bring the tears—and she showed them in her eyes—to know, as she knew by her husband's report of Anther's confidences, how poor Hawberk was taking the cruel wrong that had been done him by that wretched creature. No one else, surely, ought to insist upon justice, if *he* preferred mercy; and, certainly, if Hawberk took such a large, humane view, her husband ought to feel himself fully confirmed in it. Such a man could be trusted with the decision of what ought to be done about Hope. If he was willing to let the matter go for the present, no one else need bother.

To this conclusion, in these terms, Mrs. Enderby came; and, without transgressing the bounds of confidence in her cordiality with Hawberk, she tried to throw into her manner an appreciation, an approbation, which should be a reward to him, even in its want of relevance. As nearly as she might with self-respect, she lay in wait for him in his goings and comings to and from the mills, and she sent the very latest of her autumn flowers home by him, now to his daughter, and now to his mother-in-law, so that the old lady might not feel neglected. After one of the gay confabulations which Hawberk was as willing to hold as herself, she told him that now she knew where Hope got her happiness, and he owned that, well, yes, that sort of thing seemed to run in the family. As to his infirmity and his recovery from it, she would have liked to question him about it; but no opening offered itself, though she felt that Mr. Hawberk would have been perfectly willing to talk if they had once begun.

He was the most enthusiastic and optimistic of convalescents, and Anther, who had always to count with some sort of weakness, physical or moral, in his patients, had not the worse weakness to deal with in Hawberk. It was weakness of body, not of spirit, that confronted the physician, who could caution, but must not alarm, his patient as to his limitations. Hawberk was more strenuous than Anther in pushing their advantages against the common enemy, when he had begun sensibly to realize them. Without instruction, he suspended the laudanum altogether for a week; and one morning, at the end of it, he fell in the street, and was carried home senseless. It was just when John Langbrith had summoned his forces to the point of putting the mills into the charge of Hawberk and his business assistant, preparatory to going round the world so quickly that he would not be missed before he got back. When they told him of what had happened to Hawberk he said, "Hell!" and took up his burden again.

Hawberk went back to the alternating bane and antidote, and was much sooner at his work than John Langbrith in his scepticism could have imagined; but Langbrith's faith in him was gone, in spite of all that Anther could say or do to restore it. Even when, as the winter wore along towards the spring, and he was made to believe that Hawberk's laudanum had been gradually reduced again to nothing, and he had the witness of Hawberk's enthusiastic efficiency against his own doubts, he practised a sardonic self-denial with regard to the fact.

"You let it run along till winter," he said to Anther, "and, if he keeps up till then, it 'll be time enough to talk to me about taking a vacation. But I guess I've got enough of putting an opium-eater in charge of the mills, for one while."

In early April, when the first of the blackbirds had come prospecting as far north as Saxmills, Hawberk was one day making a personal examination of the logs in the boom at the head-gates, for certain sticks which he wished to experiment with, in a new idea of pulp which he had got. He slipped and fell into the water, still icy cold; but he easily climbed out, and hurried home, to laugh at the prophecies of his mother-in-law, who told him that he had taken his death, as soon as he came dripping into the house. For once, in a long series of gloomy forecasts, she was right. Pneumonia set in, and, twenty-four hours after it set in, death put his seal to the cure of opium-eating which Doctor Anther had effected in a typical case.

As long as she lived, the seeress could boast, not only that she knew Hawberk would die as soon as she laid eyes on him, but also that, if Doctor Anther could have attended him, Hawberk would not have died.

XXXIII

In March, John Langbrith's misery had pushed him to the desperate step of writing to his nephew that, somehow, at any risk or cost, he must get away from work for a while. It was not a case of life or death, and neither he nor Anther had pretended that it was so; but it was a case of what a man could stand and care to live. He said this to his nephew; but he said also that he had merely reached the point where he did not care what became of the business. If James Langbrith cared, he had better come home and look after it; for, in a month from the time he wrote, John Langbrith was going to leave it. Like some men who have found a grim pleasure in suppressing their feelings, and who, upon a sudden occasion, find a yet grimmer pleasure in freeing them, he poured out on his nephew the disgust he had bottled up in his heart for James Langbrith's views and aims, and said that he had better learn to make paper than plays, for more people wanted it; there was more demand even for poor paper than for poor plays. He said something about James Langbrith's being old enough to leave off being a loafer, and to turn to and do something for a living.

The letter, rightly read, was a cry of physical

pain; but there is no doubt that it was a vulgar and abusive cry, and it filled Langbrith with a fury which was not greater than his astonishment. In his whole life, his uncle had never spoken so many words to him on business, and had never offered him any criticism on what he was doing or proposing to do. He had felt a sardonic reserve in John Langbrith at their spare encounters, but so long as it continued reserve he did not care for it. He had a general contempt for his uncle, as a sort of mechanical-minded insect who could fulfil its office without volition or imagination, and now this insect had venomously risen and stung him in the tenderest part of his vanity. But he resolved to be a gentleman in repelling the attack. He determined not to answer John Langbrith's letter till he had let his wrath cool; not to judge him till he had submitted the case to another. The other was, of course, Falk, who did not give the matter too great thought when Langbrith pushed the letter peremptorily between him and a sketch Falk was making, and required to know what he thought of it. Falk read it with the sort of amusement which the pain of such a man as Langbrith is apt to give those who know him, and even those who like him; but, though he smiled, he could not refuse his friend the justice of owning, "Pretty nasty letter."

Langbrith briefly wrote back to his uncle that he was not prepared to leave Paris at the moment; but that, if John Langbrith wished to relinquish his charge of the mills, it would be entirely acceptable to have them left in the hands of his business lieutenant and of Mr. Hawberk, who, as the old and devoted friend of his father, would doubtless feel, as his father's brother seemed not to have felt, the importance and sacred character of the charge. He made no reply to John Langbrith's sarcasms, but suffered himself the expression of a high, impersonal regret that he should have always mistakenly inferred his uncle's character from his father's. He could not, however, be altogether sorry that he had credited John Langbrith with the noble nature and magnanimous ideals of Royal Langbrith. Brief as it was, the letter was as insolently foolish as it could well be, and John Langbrith, reading it on the way up to Hawberk's house, where he had been summoned by news of Hawberk's dangerous condition, pushed it into his pocket with a pleasure in not having been mistaken as to the writer which few men would have been able to feel.

He had been told that he had better go up, by the young doctor who was hopelessly looking after Hawberk in place of Dr. Anther, then in the second week of a typhoid fever. Anther had fought against the fever to the last, and when he succumbed to it he was already delirious, so that it was not known whether his asking for Mrs. Langbrith was or was not from a mind fully master of itself. But it did not matter. She was already on her way to him, at the first rumor of his sickness; and she carried her home into his homeless house, and gave him the tireless devotion in which alone she was not weak. She took her two women with her and installed them in the place, which she stripped the Langbrith homestead to make a little less comfortless. She published, so far as her action went, the fact of their affection to the whole village world. To some of those who came to offer the help she almost passionately refused, she said that Dr. Anther and she were engaged, and that they were to be married as soon as he was well again. In the sort of vehemence with which she declared this, she might well have wished to put her purpose beyond recall. Mrs. Enderby and Mrs. Garley would have helped her; there were few in the village who would not have been glad to offer help, if that of her nearest friends and his had been allowed. She was not stupidly and jealously set upon the sole charge of the sick man: it was she who had first thought of having a trained nurse from Boston, and had suggested it to the young doctor, who did not like to venture on it. She put herself second to the nurse, and subordinately shared her duties and vigils, claiming no rights and asserting no hopes they had not in common. She had not even the poor consolation of being the subject of the sick man's ravings. His crazy thoughts ran mostly upon Hawberk, whom he fancied advising and cautioning as to his case. Two or three times he dimly knew Mrs. Langbrith, but supposed himself in her own house with her. He sometimes mistook the nurse for her. All the tragedy that had allied them in the past, the baffle, the defeat, the despair was wiped out; and a trivial cheerfulness replaced it in the sick man's delirium.

John Langbrith came to tell her of Hawberk's

death, and he said to the bewilderment in which she listened, "What are you going to do about James? He ought to come home, if he ever means to; but *I* can't make him."

"I will," she said from her daze, without asking him why he could not do it, as he, perhaps, intended. But she sat still without offering to put her will into any sort of effect.

"I've got the cablegram-blank with me," John Langbrith said. "You want to cable him, don't you?"

"Yes," she answered. "What shall I say?" she asked.

"Oh, anything—just 'Hawberk dead: come immediately.'"

She wrote mechanically from his dictation; then she put in a word.

"Well," John Langbrith said, with his grim smile, "it wa'n't necessary to have the 'Mister,' but it only costs twenty-five cents more, and he didn't get the 'Mister' so often while he was alive. Want to sign it, don't you?"

"Oh yes," and she took the despatch from him. Then, after a hesitation, she signed it "Mother," and gave it back, and let him go without asking anything about Hope.

John Langbrith stayed two days for Hawberk's funeral; then, with some formality, referring to the favorable symptoms in Anther's case, which he would not have observed, perhaps, if they had been unfavorable, he broke away from his work, and took his misery with him on a vacation. He

319

had a blind notion that a sea-voyage would be the thing for him, and he thought of a trip to Bermuda. But he found that he could not get back under a week, and, desperate as he was, he could not bring himself to put that time between him and possible recall to his business cares. He devolved upon a trip to Old Point Comfort, and went and returned by the coastwise steamers, which encountered heavy weather enough to prolong both voyages, and to give him several days of haggard unrest at the beach hotel. He got in, he considered, a full week of sea-air by this means, and he arrived in New York one morning in time to take a Boston train which would connect for Saxmills, so that he could sleep at home that night.

He imagined it in this phrase before he realized, with a sardonic humor, that it would be going to bed, rather than sleeping, at home. He did not know how he was ever to sleep again anywhere; and the flame in his stomach fretted him to a white heat of exasperation with everything in life and the world. He was going back not better but worse, and he was going to take up alone the burden that Hawberk had divided with him during the last six months. Why need Hawberk have died now, damn him? He raged, and he cursed the fool for losing his life on that idiotic venture, when he could have sent any boy in the mills to pick out the right logs. In his thought he visited the insufficiency of this business lieutenant with equal fury and profanity, and wondered what hell of a muddle he would have contrived to make of things in the week that he had been left alone. He included Anther in the rage of his condemnation, for being down with typhoid just when his skill was needed to save Hawberk, and he included that young jackass of an Emering, who knew as much about practising medicine as John Langbrith knew about sailing a ship. The figure was an effect from his recent voyages, in which all forms of navigation had fallen under his contempt, as incompetent to supply a man with the seasickness on which he had counted as one of the means of relief from his dyspepsia. While the boat rolled and pitched, and cries for help hailed the stewards from every state-room, he had kept a steadfast stomach, such as it was; and he had maniacally calculated in his anguish that there was not enough water in the Atlantic Ocean to put out the fire that was burning in his hold.

It was still smoldering when the train stopped ten minutes for refreshments at New Haven, and Langbrith, who had started breakfastless from New York, recklessly decided to supply it with fresh fuel. As everything indifferently disagreed with him, he did not see why he should not have a cup of turbid coffee, a plate of cold beans, and a piece of apple-pie, as well as anything wholesome, and he was wiping the traces of this repast from his shaggy mustache when he ran for his train, and scrambled into his parlor-car, just before the porter picked up his carpeted step and swung himself aboard. As he crowded through the narrow aisle on his way to take his seat again, he glanced into the smoking-room and met the eye of his nephew, who turned at the same moment from watching the shipping in the harbor through the windows and over the platforms of the cars receding on the sidings.

They knew each other with less surprise on John Langbrith's part than James Langbrith's; but it was the uncle who expressed an ironical astonishment, when he decided to be first to break the silence in which they were glaring at each other. "Oh!" he said, "*thought* you'd come over!"

XXXIV

EVERYTHING in the sight of the young man made the older man hate him; but, most of all, it was the indefinable touch of Europe, of France, of the Latin Quarter in James Langbrith's dress which, while it could not interpret itself explicitly to John Langbrith's ignorance, expressed something superiorly and offensively alien.

"Uncle John"—the young man's misfortune was to intensify this effect by the tone of his suggestion—"don't you think we had better leave anything of this sort till after—till later?"

"No, I don't," John Langbrith sourly responded. And he came into the smoking-room, and sat down in a chair opposite the corner of the sofa where James had been looking out of the window.

They had the place to themselves. It was the train which used to be called the "ladies' train," because of its convenient hours and slower gait, suitable to the leisurely transit of the unbusiness sex; and James Langbrith, in entering the car, had noted that, but for one man, there were only women in it, and had taken possession of the smoking-room to think the more unmolestedly of things that had filled, it seemed almost to bursting, his mind for the last ten days. John Langbrith had made no such

323

observation, but he saw that they were alone with an opportunity for quarrel, with which he luxuriously toyed before he fully grasped it.

"When did you come?" he asked, after looking vainly for a splinter to chew upon. He caught sight of the porter's whisk-broom over the wash-bowl, and supplied himself with a straw.

In the mean time, James had said, "We got in this morning; our boat was thirty-six hours late; it was two days before I could get away after the cable reached me. She was the first boat out."

The words were spare enough, but there was an exculpatory flavor in them that suited John Langbrith's ferocious mood, and when James added, "How is my mother, and Hope?" he loosed himself upon the young man.

"I don't know. I haven't seen them for a week, and I don't want to bandy any small-talk with you. I got your answer to my letter all right, and I want to have a square understanding with you. I don't know as we ever had a regular understanding, did we?"

"I don't know that we did, if you mean about the mills."

"I mean about the mills. What the devil else could I mean?"

"That," said James, "was all arranged before I was old enough to have any understanding with you, and since then I have let my absolute trust in you take the place of an understanding."

"I know that damn well. But the time has come now when I don't want your absolute trust."

It occurred again to James Langbrith, as it had occurred before, since getting his uncle's astounding letter, that his uncle might be mad.

"I want to know whether you've come home for good, to take a grown man's share in your own business?"

"That depends," James parried the issue. He was really no more afraid of the impending quarrel than his uncle, but he was a dreamer, and he liked to nurse his conclusions before trying them: liked to shy off from them and feign that they were not immediate, and perhaps not at all. John Langbrith was concrete where the young man was abstract, and his pleasure was to force the issue.

"It don't depend on me. I'm done with the thing. I'm going back to Saxmills, but it's to pull out for good and all."

"I suppose," James Langbrith assented, "that there will be an accounting and a settlement?"

"Oh, don't you be afraid of that, young man. There'll be a settlement all right, and after I've been paid a little more than days' wages, you can have the rest." John Langbrith felt the coffee and beans and pie beginning to ignite, and he flamed out upon his nephew from that inner conflagration, "What do you mean by 'an accounting,' you—you whipper-snapper?"

James Langbrith made no answer, and his uncle pulled his chair closer, and put his face so near that the young man turned his own slightly aside, to get it out of the current of his uncle's dyspeptic breath.

"What do you mean? What do you mean?" John Langbrith insisted. "Do you suppose Royal Langbrith was a man to put anybody slippery into his business?"

"You know," James Langbrith disgustedly, but quietly, responded, "that I could not mean to impugn your honor." With the effect of being pushed to the wall and menaced there, he looked like his mother, who had so often been in that place, first through his father's duress and then through his own.

"Honor!" John Langbrith spat the word out of his mouth. "*I'm* talking *business!* What sort of man do you suppose your father was, anyway?"

A light, less of hate for his uncle than of love for his ideal of the father he had never known, kindled in James Langbrith's eyes, the long eyes of his mother. "He was, at least, a gentleman."

"That's to say I'm *not*. Well, go on! We'll take it for granted in my case. How do you know he was a gentleman, heigh?" He pressed him with the last word, and repeated it with a smile of scorn and pain. "Heigh? How do you know?" James Langbrith moved his head from side to side, as much now to escape what message of disaster might be coming as the effluvium that should bear it. But he made no answer, and John Langbrith hitched himself so near that his bony shins sawed against his nephew's legs, and he tapped him on the knee with his spiky forefinger, in the habit he had when talking business with people. He was talking business now as he said: "You don't know? Well, I do, because he was my brother, and I knew him up to within twenty minutes of his death. If he didn't reform within them twenty minutes"—John Langbrith in his passion lost the grip, always uncertain, of his grammar—"he'd ought to have went smack, smooth to hell, like shot out of a shovel!" James Langbrith's eyes dilated with the assured conviction of his uncle's insanity, but at the same time his nostrils swelled with resentment of the maniac's offence. John Langbrith gave him no chance for the expression of either the belief or the emotion. "Ever since I could remember him he was the coolest and slickest devil! I don't know where he got it! He had the trick of making other folks do his dirty work—and he was full of that, I can tell you—and keeping such a hold of 'em that they never had the chance to squirm out of the blame. He had *me* fixed good and fast, while we were boys, by a scrape he hauled me into along with him, and when he wanted me, any time, and said 'Come!' you bet I went. That's the way I came to be left in charge of his business when he died, and that poor fool of a Hawberk crowded out of it with lies that Royal threatened to tell his wife if he peeped. That's the way the woman Royal lived with down to Boston came to take what he give her and no questions asked, without makin' trouble for him, alive or dead. *She* was fixed so that *she* didn't peep! And so right along the whole line! If he hadn't cowed your mother for good and all she might have said something about the way he used to bully her, and when he came home from his Boston sprees used to pound her.

Oh, he was a *gentleman*, Royal was! And that poor sheep of an Anther might have spoke out in meetin' if your mother hadn't been so mollycoddlin' about you that she couldn't bear to have you told the truth when he wanted to marry her and couldn't make her tell. But *I'll* tell you now, and don't you forget it. Royal was such a gentleman that he cooked it up with the devil how to fool the whole town, and make 'em believe he was a saint upon earth. That library buildin'! He gave it out of the profits of the first year after he choused Hawberk, and the mis'ble crittur was makin' it all right for Royal by tryin' to kill himself with laudanum! Why, he made Royal Langbrith rich with his inventions, and then Royal got the credit of 'em; and he got the credit of doin' the handsome thing by a man that was an opium-fiend, according to his tell, from the beginning. And when you took it into your fool head to put up that tablet to him in the front of the library, he had things so solid that all hell couldn't bust 'em up. Anther did go round to Garley and tell him the rights of it, but that old chump honeyfugled him into believin' that he better let by-gones be by-gones, for fear of the corruptin' effects on the community. Then Anther come to me, the last thing, but I was stickin' to my job, just about then, and I thought if your mother wouldn't keep you from runnin' your neck into the noose, *I* wouldn't. I believe there wasn't a last one of them jackasses up on the platform that wasn't as big a fool as you, except me and Anther, and that old honeyfugler. And I ain't sure," John Langbrith said, withdrawing his furious face a little from its proximity to his nephew's, "but what I'd have held my tongue, now, if you hadn't put it to me that Royal Langbrith was a gentleman and I wasn't; but now you've got it, I guess, about as strong as they make it, right in the collar-button!" He leaned forward again, and demanded in a fresh burst of fury: "I suppose you don't believe me! I presume you think I'm tryin' to work you, or off my nut, or just pure ugly! Well, you can ask Anther, when you get home. And you can ask your mother! And you can ask the mother of his children — I'll give you her address. And you can ask that old honeyfugling fraud of a Garley. And you can ask Haw— Oh no, you can't ask him! He's out of it, but I guess his mother-in-law could tell you something she's suspected, all right! Oh, you've got a nice job cut out for you, young man! Why, I wouldn't be in your shoes—"

The parlor-car conductor put his head in at the door, and looked at them. John Langbrith fell suddenly as silent as James Langbrith had been throughout. With the shadow of a changing mind passing over his face, the conductor said, "See: d' I get your tickets?" and James Langbrith, if not John Langbrith, knew that he had been drawn to them by the sound of a noisy, angry voice, and had meant to ask them to be quieter.

But the young man could not care. It would not have mattered to him now whether the whole world had overheard; the universal knowledge of the fact could be nothing, compared with the fact itself.

His uncle got up and went out to his seat in the parlor, but James Langbrith did not move. He sat exposed to the tempest that had opened upon him without the shelter of a doubt. It seemed still to rage upon him like some war of the elements, and he was aware not only of the truth of what had been told him, but of its not being novel. He had that mystical sense of its having all happened before, long ago, and of a privity to it, in his inmost, dating back to his first consciousness. The awful conviction of the reality which held him like a demoniacal obsession was blended with a physical loathing of his uncle's person, a disgust verging on sickness for his boiling hate, his vulgar profanities, mixed with the oldest and the newest slang, and the brute solecisms of the vernacular into which John Langbrith had lapsed in his passion. If he had wanted proof of what had been said of his father, the fact that John Langbrith was his father's brother would have been proof enough to the young man's shame.

From time to time, in the turmoil of his cognitions, he had a nerveless impulse to follow his uncle, where he had gone to his seat in the drawing-room, and ask him this and that, but he did not. He was not aware of stirring till the porter came for his bag at the South Terminals in Boston. Then the horrible dream went on like waking, as he drove across the city to the Northern Stations, and found his train for Saxmills. Till then he had lost sight of his uncle, but he saw him boarding the same train; he looked into the smoker, and, finding it fairly full, he got into it, making sure that John Langbrith would not come to molest him there. He had no wish now but to keep away from him, to keep for the present out of the sight of the man who had heaped his dishonor upon him, and who alone of all that he could encounter would be knowing to it.

Apparently John Langbrith had no wish to look him up. He had doubtless poured the last drop from the vials of his wrath out upon him, and was without any purpose of breaking them upon his devoted head. At any rate, when they got out of the train at Saxmills, the uncle made no motion to approach his nephew. He stared at him, ignoring him as perfectly as if he were any other shadow of the vaguely lighted depot, and getting into one of the two ramshackle public carriages which had chanced a late passenger, drove off into the darkness. James Langbrith took the other, and bade the man, who was a stranger to him, drive to Mrs. Langbrith's.

All the way he had a sinking of the heart which was not related to the failure of his mother to have him met, after he had telegraphed her from New York that he was coming on that train. There was no lifting at sight of a belated lamp in the parlor, or at its moving thence, when he knocked, and showing through the transom of the hall-door, which his mother opened to him herself.

XXXV

James Langbrith took his mother in his arms with an emotion that he had never known before, with pity, with honor, with reverence due to mute suffering, with everything that endears and exalts an object long beloved and wronged. She seemed surprised at his warmth, and sparely kissed him, without even a lax return of his embrace.

"Mother," he said, breaking from the sense of her coldness and from the subjective pressure of something unwonted in the absolutely unchanged environment, "I came from New York with Uncle John, and he told me about father." As he said this, he noted that the place was lighted only by a hand-lamp, which she was nervously fingering. Her face was swollen as with weeping, and the red crescents under her eyes were tumid with tears unshed.

She said, beginning with the estrayal of his glance towards the lamp: "Norah is not here, and I have let the cook go to bed. I said I would sit up for you. *She* wanted to."

"Thank you," he said, mechanically, to her drooping head. "Uncle John," he repeated, "told me about father." Either she did not understand or she did not heed; it seemed impossible that she should not have done both; but he felt that it would

332

be cruel to press her further with the fact of his knowledge now; he took his first lesson in forbearance with her. "I want to see Dr. Anther, at once. Do you suppose he is well enough to see me, to-night?"

"Dr. Anther?" she asked, with an accent that impressed him as having something in it as strange to herself as to him. "Why, you can't see him!"

"Yes, I know he is sick; Hope wrote to me. I didn't think—you must excuse— How is he?"

"He is dead," she answered, simply. "He died early this morning. I wanted to stay and sit up, to-night, but they wouldn't let me. They say it isn't the custom, any more. I've just got back here. I brought the trained nurse. She ought to have a little rest before she goes back to Boston." She added one fact to the other in the same quality of tone, with the same effect of not realizing any of them.

"Dead?" was all that James Langbrith could say.

"They thought he was getting well, one while; or I did. But Dr. Emering said he was afraid, all along. He had splendid care. That trained nurse is as good as another doctor." With the same lifelessness she said: "I've put you out a little supper; and then I suppose you'll want to go to bed. I don't know as you'll find things very comfortable. I took both the girls with me, and, with Norah there still, things haven't been put to rights, all. But I've got your room ready."

She ceased to speak, and they both sat in a silence like that of the night when he found her in the moonlight there after his return to do Hope's bidding, and consent to her marriage with Dr. Anther. Now as then it was as if there was to be no end to their sitting in silence together, but now it ought to be a silence that united, not parted, them.

Up to a certain moment in every evil predicament men are the victims of it, and after that, if they continue in it they are its agents, though as little its masters as before. They are exceptionally happy men if they realize this early enough in life to make choice of their better selves against their worse, and in that choice finally prevail over their evil predicament. The events of James Langbrith's situation presented themselves with the simultaneity with which events are said to show themselves in instants of mortal peril. No detail was missing in the retrospect of wilful arrogance, of blind conceit, of vain folly, of baseless illusion; and yet, with it all, he justly felt that he was not so bad as any of the things he had done. At his age he could not be without hope: there could be as yet no error in life wholly irreparable. His soul seized upon renunciation, sacrifice, as its only refuge, and he said, as he thought, to himself—but from her response he knew that he must have also said it to his mother—"I must release Hope."

She answered simply, "It's too late, to-night."

"Yes, but I will see her the first thing in the morning, and tell her. That will be the end between us." His mother did not gainsay him, and he asked: "Does she know about it—what my father did to hers?"

His mother said impassively, "I don't believe she does."

"Then I must tell her, and let her take herself back. She would hate me."

His mother looked at him in a daze; she seemed about to speak, but did not. "Mother," his voice quivered in the question, "do you suppose Dr. Anther hated me?"

She took time, as if to consider. "I don't believe he did—after the first—after you went away that day. As far as anything went that he said then or ever afterwards, he pitied you."

"Oh!" Langbrith groaned.

"I don't know," she resumed, "how much for me it was that he pitied you. He was always wanting you to be told about—about Mr. Langbrith; but he wouldn't force me, when he saw I couldn't. I don't know as I did right not to tell you, but the time never seemed to come."

The words had a sound of excuse, and against this he protested, "Oh, mother!"

"He wanted me," she continued emotionlessly, "to let him tell you, but he always said he wouldn't be my tyrant; he thought I had had enough of tyrants."

Her son winced with a cruel pang. "Did he think I had been your tyrant?"

"I guess he did, in some ways. But not that you meant to. He never liked to blame, a great deal." She added, with finality, "He was a good man."

"Yes, yes!" Langbrith wailed, in his intolerable

regret. "He was a good man. And I insulted and outraged him when, because he meant the best a man could and had been your true and constant friend, I should have been on my knees to him. And mother, do you?"

"Do I what?"

"Pity me, too? Forgive me?"

She drew a long, weary sigh. "Oh, what does it all matter?"

"Everything—the whole world, life, death!"

She appeared to consider again. Then she answered, "I don't know as I ever felt but the one way to you. You were my son."

He felt that to rise and kiss her for the assurance of her love would have been to profane it. He sat where he was, but he burst into a wild sobbing, the tears of a man who does not weep till the fountains of being are broken up. When he controlled himself he asked, "Who else knows about father?"

"Dr. Anther said he told Judge Garley and Mr. Enderby. I shouldn't be surprised if Mrs. Enderby knows too, but I don't believe Mrs. Garley does. Mr. Hawberk did. And your Uncle John. I guess that's all."

"And now everybody must know! I will begin with Hope."

His mother said nothing to this; it was as if she considered it his affair, in which she had no longer any part. She sat awhile, but not apparently for further speech with him. Then she rose and took her lamp. "I guess I will go to bed, now." She moved absently towards the door. She turned, and came back to light another lamp, which stood ready on a table. "I was leaving you in the dark—"

"I would rather," he broke out. "Don't light it! I can find my way. Good-night, mother!"

She looked at him, faltering, and then she stooped and kissed him on the forehead, and left him sitting in the dark. He realized that he was sitting before his father's portrait, and that it had been witness of the scene which had passed. He mutely said to it, "I must begin to undo."

He sat through the night, and in the morning, Norah returning to the house, and letting herself in with a latch-key at the front door, woke him from the drowse he had fallen into, and after his bath forced him to drink the coffee she had brought him in the dining-room. She was very gentle with him, and he with her, like people sharing the sorrow of the same house of mourning, but beyond the exchange of a few questions and answers about his voyage home they did not speak till he said, "What did my mother mean, Norah, about having just got back here? Has she been out of the house?"

"And didn't she tell you? We all been up at the doctor's keepin' house there, and doin' for him; me and Mary and your mother, ever since it was sure he was goin' to be bad. I thought some one would be writin' to you!"

"No," Langbrith answered, briefly.

"Miss Hope was with us, too, some of the time, and Mrs. Enderby. But it was all no use, as far as the doctor went. He didn't know one from another, after the first day or two. Mary has got

337

ye some rice - cakes, Mr. James. Won't ye have anny?"

Langbrith was pushing back his chair. "No, I don't want anything more, Norah. I'll be back before long; tell my mother, when she comes down."

"And I hope, then, she won't come down soon, if she's sleeping. It's more than she's done for the last week."

He went away with the trivial sense of Norah's Yankee correction in her Irish parlance, which he did not remember to have noted before, and he had no question of going directly to find Hope at eight o'clock in the morning.

She was waiting for him, even then, though it could not be said that she was expecting him. He had figured holding himself from her out of duty to her, but they were in each other's arms before he could help it. In that mutual transport, and while he still pressed her close to him, she divined his constraint, and asked, vividly, "What is the matter?"

"I want to tell you, but I don't know how," he began.

"Well, don't *mind* now," she said, with the first gleam of her inextinguishable gayety. "Do it anyhow," she added. "There isn't anything I can't bear now—now *you're* here."

"Oh, Hope, dearest!"

"Is it something dreadful? something about *us?*"

"It's about your father"—she pulled herself away, he felt indignantly—"and mine. I should think I was dreaming, but I know I'm awake for the first

338

time in my life. Every one must know the truth,
but I must begin with you."

"What do you mean, James Langbrith?" she
demanded, severely. And he found the strength of
despair.

"My father was not what I believed. He was a
man that—that—wronged every one he had to do
with. He wronged your father so cruelly that he
drove him to the opium."

"*Your* father? *Mine?* Why, you must be crazy!"

"If you say that you will make me so. But I am
perfectly sane at last. Uncle John told me about
it yesterday coming up from New York, and I've
come the first thing this morning to tell you. I
told mother last night that I was coming to release
you, and to give back all that my father had stolen
—stolen!—from yours. It makes me feel as if I had
stolen you."

"Now, James Langbrith," she broke out upon him
from her bewilderment, "you just stop being silly,
and tell me exactly what you're talking about."
She took his hand, and pulled it vehemently while
she fixed him with her eyes.

He began again, and now he told her the greater
part of the story that John Langbrith had vindic-
tively poured out upon him. He could not bring
himself to speak of his father's hidden life; the innocent shame that was between them forbade that;
but, somehow, he possessed her of all else that he
knew, while she kept clutching his hand convul-
sively, and pulling herself to him. "This has been
my home-coming. I—didn't sleep last night, and

I'm rather broken up, or else I could have prepared you—"

"Oh, you poor thing!" She put forward her left hand and passed it over his reeking forehead, as if he were her child, in the divine mother-pity which is in a woman's heart even for her husband or her lover. "You are the injured one, kept in the dark so, all your life."

He tried to resist her compassion, but his head fell upon her breast. "It had to be so. And now," he said, "the most I can do is to make restitution of what you have been robbed of, and give you back yourself."

"Oh, how ridiculous!" she said, with a bewitching inadequacy, while she smoothed his hair with her hand. "Do you suppose father would want *you* to do that? And I won't have myself back, as you call it! What would I do with myself, if I had it?" she added. "Now you be still, and let *me* talk awhile. I don't believe it's as bad as your Uncle John says, and, if it is, it don't make any difference now. It's all past and gone, isn't it? I guess father got the fun out of his inventions, even if somebody else got the money. He was so happy this last year that it would have made up for anything. I do believe that he couldn't have enjoyed it so much if it hadn't been for what went before. He never said a word to me to show that he felt injured, and he liked *you*, James; he was proud of you, and he believed in what you were trying to do, over there, even when I couldn't, always. Father was a genius, *I* think. Don't you?"

"Yes—"

"Well, then, he had his good time as it went along. He took it with him, as you may say. And as far as I'm concerned, and that restitution of me that you talk about, I guess we'll just have me in the family."

If his despair had been what he thought it, he could not have resisted her sweetness, her greatness; he could not have denied himself the pardon and the blessing it assured him. But he could not speak, and a little hurt at his silence stole into her drolling voice.

"Still, if you don't want me—"

"Oh, my dearest!" he cried out. "What are you saying?" and once more they took each other into a long embrace that said everything which they had both vainly tried to put into words. When they were so far parted that he could look into her eyes, he said, "How strange you are, Hope!"

"Am I? Well, that's what Dr. Anther used to insinuate, so it's a compliment that I'm used to. He seemed to think it was all right, even if you don't."

"I? Oh, Hope!"

"Well, *some* people, then. If they were in your place, they would say that it was very queer I shouldn't act more as if I felt father's going. And we haven't spoken of it; poor father! What would you say if I said sometimes I was glad of it? He was well when he went, and he hadn't touched a drop of laudanum for months and months. But I never felt sure about it, and I don't believe Dr.

341

Anther did, and when I think how he used to suffer
—well!" She was one of the women who rain and
shine together, and now the tears fell over her
pathetic smile.

"I know," he gulped.

"Sometime I'll tell you all about him, but not
now. And I'll tell you about Dr. Anther. He
was the best man that ever lived. Are you glad
that you went home that night and took it back,
with your mother?"

"It's what gives me the only courage I have
left."

"Well, I'd rather hear you say that than that *I*
gave you courage," she said; but he could see that
she was a little jealous of the help of even a good
conscience, and he answered, "You're my Hope."

She laughed into a sob, and then laughed out of
it. "Then you must be equal to seeing grandmother. Come in and speak with her."

They had been sitting in the dim little parlor,
and now Hope led him into the dining-room, where
Mrs. Southfield was grimly chastizing the breakfast-
table for the disorder in which Hope had left it
when she flew to let Langbrith in at the front door.
She paused with a plate in her hand, and transferred
her fierceness to Langbrith's face. "Here's James,
grandmother," said Hope, recklessly; "can't you
stop and shake hands with him?"

"I don't know," the old woman said, "as I want
to shake hands with any of his tribe."

"Not when he's going to be one of our tribe,
grandmother? That's what he says he is?"

"I wouldn't trust anything a Langbrith says,' Mrs. Southfield returned, with impartiality.

"Well, then, it's what I say, too. Just shake hands, anyway," Hope bade her cheerfully, and, after her grandmother had wiped her hand on her apron and given it to Langbrith, the girl pursued, "Well, now, that's settled"; and when she had drawn him out of the room again by the hand that was still finding itself in his, she suddenly asked him, "Did you like it?" and at his stare she added, "The way grandmother welcomed you?"

"It was what I deserved," he answered, stonily.

"No, it wasn't, but it's what you'll get if you tell everybody about your father. Will you do it? Can you?"

"I *will*, whether I can or not."

"I don't like that hard look in your face," she said, with a criticism that seemed general rather than special; then, with special application, she said, "It makes me afraid of you. I wonder if you'll be stubborn."

"Don't you want me to be firm in the right?"

"Yes," she sighed, "if you know what the right is."

He looked at her, perplexed. "Have you told any one else?—or no, you said you wanted to tell me first. Are you going to tell other people right away?"

"Can it be known too soon?" he demanded, gloomily. "I should like to stand by Dr. Anther's open grave and proclaim it, and take my father's shame on me before them all."

343

She only said, "Oh!" with so little liking for the imaginary spectacle that he had to brace himself for the effort of going on.

"That tablet must come down out of the library as publicly as I put it there. I must tell the whole community the facts of my father's life, so far as they can be decently known. I must own the wrongs he did, and ask any man who has a grievance against him to come forward and let me right him so far as I can."

"It sounds like a play, doesn't it?" she said, with a smile that was somehow loving as well as mocking. "Anybody can see that you will know how to write plays." At sight of the dismay in his face, she turned wholly serious. "James, you are crazy! Don't you see that it wouldn't do?"

"Why not?" he faltered.

"Because it is too late! You would just disgrace yourself and not help anybody. It would make the greatest scandal! And what good would it do?"

"That is not the question."

"Yes it is, James; and if we are going to bear this together—"

"What have you to do with it?"

"Well, if I don't *take myself back*, I should say I had full as much to do with it as you!"

He stood daunted by what had not occurred to him before, and he could not answer her anything.

"Now do you understand?" she triumphed, tenderly. "I guess if it was my father that suffered the most I have the right to say the most; and I don't believe I should like to have everybody know the

344

kind of family I was marrying into. Why, if grandmother treats you the way she does because she felt it in her bones about your father, what would she do when all the neighbors knew it, and it got into the papers? Think what Jessamy Colebridge would say; and Susie Johns!"

He knew that she was entreating him lovingly as well as mockingly, and though it was sweet, yet he could not make sure of the reality of what was so opposite to the picture he had carried night-long in his mind of her instantly agreeing with him, and supporting him in the ordeal he proposed to himself, in the event of her refusing his renunciation. "I don't understand you, Hope," he hesitated.

"Yes, you do, James Langbrith!" she retorted. "You see that I've got just as much to do with this as you have. Don't you suppose," she softly reproached him, "that I know how you feel, and how proud I am of you for it? But I'm not sure about it—I'm not sure it's right; and I'm not going to let you do it on your own responsibility, if I have any say in it. And I have, haven't I?"

"Why, surely! If I hadn't been so blindly selfish I should have seen that without your telling me."

"I will settle it about your selfishness some other time. It's *my* selfishness now. This is something we can't decide between us. Do you know what I was just thinking?"

"Yes," he huskily responded. "That we could leave it to Dr. Anther."

"Yes!" she said, solemnly.

"I am glad you knew. Who else is there?"

"My mother—"

"We mustn't put anything on her. But she had a right that you should think of her. Well?"

"Uncle John would be no use."

"No."

"Judge Garley?"

"Of course you don't mean it. He is a good man, but he would just laugh at us. Why are we beating about the bush so? We must go to Dr. Enderby!"

"Yes, I really thought of him next, when I remembered that Dr. Anther—"

"I knew you did. Well, we ought to go to him at once. Don't let us hesitate. Wait till I get my hat."

She went up the cramped stairs, apparently into that chamber out of which he had once heard the nightmare groans of her father coming, and before she returned he heard her open some door downstairs, and call cheerfully through it, "Don't you wash the dishes, grandma. I'll be back soon," and she joined him with her face freshened and brightened by the bathing away of her tears.

Her quick tilting was swifter than his long striding as they descended the hill-side path towards the rectory, and she chanted to Mrs. Enderby among the flowers beyond the fence with a gayety that she could not quite keep out of her voice, "How d' ye do, Mrs. Enderby! Is Dr. Enderby at home?"

XXXVI

THE two young people were upon Mrs. Enderby before she could drop her garden-shears and dismiss from her consciousness a prescience of their coming for a purpose she had long associated with them and replace it with a decorous sense of all there was in the circumstances of their lives to banish that from them for the time. She was smiling too radiantly upon Langbrith, she felt, even when she had effected the substitution, but she could not help it. She could only make an apposite reflection on the strangeness of life as she asked him about himself and about his mother, and dedicated some just observations on the sad home-coming this must be for him in the losses which he shared with them all. Then she said, "The doctor is in his study. Won't you go in?" and offered to remain outside; but Hope said:

"We want you, too, Mrs. Enderby. It's something that we want you both to talk with us about; don't we, James?" she ended, with a deference to him which seemed to Mrs. Enderby very pretty.

"He is trying to write his sermon—for to-morrow, you know," she explained more directly towards Hope; but it was now Langbrith who answered:

347

"If it is the funeral sermon, what we may have to say will be fit, perhaps."

"Oh, he will not mind being interrupted by you, in any case," she said, with her mind playing mechanically away from the occasion to the general duty she had of always sequestering the rector when he was writing.

After the greeting to Hope and the formalities with himself, Langbrith took the word with a dignity and composure that Mrs. Enderby saw kindle the girl's eyes with pride in him.

"I was saying to Mrs. Enderby that I hoped our errand wouldn't be out of keeping with the subject of your sermon, if you are writing about Dr. Anther. He knew something — something of my — family history which never came to me till yesterday. My ignorance of it was the means of a cruel misconception on my part and of most generous forbearance on his; and it is a question now of what can be done in reparation from me — the sort and measure of it."

Langbrith paused, and the rector sat kindly interpreting the young man's thoughts by the light of his previous knowledge. But it was not for him to forestall the confidence which he felt was about to be offered to him. He merely said, "I could hardly imagine anything you could tell me that would heighten my sense of Dr. Anther's worth."

"Yes, I know that," the young man assented, with a humility which made the other accuse himself of having not been quite clear. "But before I speak of him, I ought to say that I owe *you* some

348

reparation. When I asked you to say some words at the dedication of the tablet to my father, I didn't know that my father — that my father —" He choked. He had easily told Hope, not only because, as she had made him realize, it was as essentially her affair as his, but because, also, there was something in the confession of his father's iniquity to one so supremely concerned which supported him; but his heart sank with a sense of the common shame awaiting him from the common knowledge, as it intimated itself to him from even such pity as Dr. Enderby's. He perceived that it was not the victims of his father's misdeeds that he feared, but the witnesses of these whom his confession would create. Instinctively, he looked towards Hope for help, but she dropped her face, and at the pathos of this Mrs. Enderby addressed a murmur of appeal to her husband.

Probably he saw no reason for putting Langbrith to the ordeal he shrank from, and he said: "You needn't go on. I think I know what you want to say. I did not know it when you asked me to speak those words, but I knew it before I spoke them — from Dr. Anther."

Langbrith fetched a sigh of relief that was almost a groan. "I won't say," the rector continued, "what I might have done if I had known it all when you asked me, for I am no longer master of such a situation, and I can't go back to it and re-create it. But I was informed in time to refuse a part in that ceremony, and I did not, for reasons that still seem to me good."

Langbrith passed his right hand over his fore-head, and was aware of having Hope's hand in his left as he did so. "Would you mind," he huskily asked, "telling me your reasons?"

"They were not very profound. They related less to myself than to the effect of my refusal with the public — of the ultimate effect, if the cause of my refusal became, we will say, notorious. I had not much time to give to the matter, but I find that I don't think differently now, upon further reflection. It seemed to me that no good and much harm could come of revealing the past; that so far as your father was concerned we had no right to enter into judg-ment, and that so far as God's purposes were con-cerned we had no right to act upon our conception of what they might be in such a case. Do I make myself understood?"

"Yes," Langbrith whispered.

"I believe that I said to Dr. Anther—in fact, I am sure I did—that to take upon ourselves any agency for supposed justice—for the discovery and the retribution implied by the concealment and the wrong in the case, would be in a manner forcing God's purposes; I don't like the phrase, now, but it expressed my meaning. May I ask how the matter has become known to you?"

"My uncle John told me yesterday, as we were coming up from New York. We have had a differ-ence about — the business, and I am afraid I — I affronted him; and—and he told me."

"In anger?"

"Yes, in anger."

The rector thought how it was written, "Surely the wrath of man shall praise Thee." It seemed to him that the Divine Providence had not acted inopportunely; and he was contented with the mode in which the young man had learned the worst; it was better that he should have come by the knowledge of it so than by any deliberate revelation, with the effect of such authority as an officious interference could have arrogated to itself. His mother could not have told him, and she could not suffer Dr. Anther to tell him; but his father's brother might tell him, in anger and in hate, even, and out of his evil passions, and the evil passions they would arouse in the young man evoke the best result possible from the otherwise hopeless case.

Langbrith waited for him to speak; then he said: "And what do you think I ought to do now?"

"Oh, I beg your pardon. What had you thought of doing?"

"Of making it all known; of undoing my father's wrong as far as I could, and of revoking my own acts in perpetuating his good name—the good name he has falsely borne in this community."

"That is natural—for you, and you will let me say that it does you honor. But— What do you think, Hope?"

"I think he oughtn't to do any such thing."

"Why?"

"Because I don't see what good it would do, and it would make a great deal of misery for nothing. I know that the Bible says things have got to come out, but it doesn't say that they need come out *here*, when there's nobody left to suffer for them but those that didn't do them."

"What do you say, my dear?" The rector turned his head towards Mrs. Enderby.

"I say what Hope does." Mrs. Enderby's eyes shone with admiration of the girl, as she smiled on her.

"And I suppose there can be no doubt of your mother's wish?" he asked Langbrith.

"I am afraid," said the young man sadly, "I hadn't considered her. I'm afraid that I have never considered her."

The rector sat in a muse which he was some time in breaking. "If it is something that you feel is for the good of your own soul," he spoke solemnly, "I could adjure you to speak out and make confession of your father's sins."

"I was trying," said Langbrith, "not to think of my own good." He looked at Hope, as if there might be some help in her, but she would not meet his glance.

"Then," said the rector, "though I know that it would be a relief to you to have all this known, and to take upon yourself the dishonor which the stupid and malignant love to visit upon the children of wrong-doers, I think you must not seek that relief. I would impose a more difficult, a heavier penance. I would bid you keep all this to yourself, as your mother has kept it to herself, and as your wife— Excuse me, I didn't realize—"

"Oh, that is all right, Dr. Enderby!" Hope quaintly condoned his break.

"—and as your wife," the rector resumed with fresh courage, "wishes you to keep it. I know that from my talks with Dr. Anther; this was finally his mind in regard to the matter, and he told me this was finally, or indeed, long ago, the mind of Hope's father. Yes, you must keep this secret locked in your own heart, until such time as the Infinite Mercy, which is the Infinite Justice, shall choose to free you of it. You will know the will of God when, if ever in this world, there is some event which may well seem a chance, leading to the discovery of what you have kept hid. Then you must own the truth promptly and fully. I believe in your good-will, and in your love of the truth, and I know that God will give you strength to do His purpose when He bids you."

XXXVII

At Anther's grave, Enderby kept himself to the ritual of his church, and disappointed many who thought he would make some remarks, as they phrased it, on the dead man's life, more final than anything he had said in his sermon the day before. There was some disappointment with the sermon itself, which the rector shared, for in his reluctance to make it the mere personal praise of his friend, he was aware of having kept it too general. He would have agreed, if he could from his own knowledge, with those who said it was the least moving of the discourses of the day which had all dealt with Anther's character and career. At the Orthodox Church, the Catholic Church, the Methodist Church, and the Universalist Church, the qualities of the man who had now become a memory were dealt with directly, and his example interpreted as a lesson to those who heard. But Enderby shrank from eulogy, and while he knew that he was failing the expectations of his hearers, he had the consolation, such as it was, of knowing that he was dealing with Anther's memory as Anther would have had him if it had been his to choose. Even this consolation was alloyed by the consciousness that it was no more for Anther to choose being made little of than to choose being made much of, and that in deferring to an imaginable preference of the man he was possibly as greatly in error as if he had pronounced the warmest and fullest panegyric of his virtues.

He could only say to himself that he had done what he could, when he feared, from the effect, that he had not done enough. He was curiously disabled by the personal considerations of the case, not only as concerned Anther himself, but as concerned Langbrith and his mother. In the friendship beginning tardily, but growing rapidly into something vitally strong between them, Anther had told the preacher of all that had passed with either of these and himself. He spoke of the affair as if it were a great while ago, and with a certain aloofness in which he judged himself as impartially as the others. From being the man in later middle life who had wished to form the happiness of a woman long dear to him, he had suddenly lapsed into an elderly man to whom it was appreciable that he could not have made her happy, but only more miserable, if he had pressed her to obey the prompting of her own affection for him. He had come to see that in a case where nothing was wrong, where everything was right, there were yet obstacles which could not be removed without a violence leaving a bruise destined to be lastingly sensitive. In his confidences to the man who understood him, he not only excused James Langbrith's part in the matter, as something natural and inevitable, but his tolerance retroacted towards the boy's father, and he accounted for Royal Langbrith with a scientific

largeness in which Enderby could not join him.
He seemed to have exhausted the hoarded abhorrence with which he had hitherto visited the sinner's
memory, and to regard his evil life as a morbid condition with which the psychological side of pathol-
ogy had to do rather than morals. He regarded him,
apparently, with no more resentment than some
treacherous and cruel beast whose propensities imply
its prey, and which has satisfied them with a moral
responsibility difficult or impossible for our ethics
to adjust. In these speculations, Royal Langbrith
seemed for him a part of the vast sum of evil, not
personally detachable and punishable. As for that
publicity which his revolted instincts had long demanded for Langbrith's sins, he divined that it
would have been the wildest and wantonest of
errors. He alleged the attitude of Hawberk towards the memory of his pitiless enemy. Hawberk
once said that he guessed Royal Langbrith was
built that way, and that it was too late to give him
a realizing sense that there was anything out of
order in his machinery. Hawberk said he had no
wish to make anybody else suffer for what Royal,
as he called him, had done. He doubted whether,
if Royal himself were on hand, he should want to
collect anything from Royal out of his pocket or
out of his hide. He guessed his claims were outlawed.

Anther himself more than once approved the
position which Enderby had taken in regard to the
public celebration of Langbrith's public munificence,
but in this he did not allay the disquiet of the rec-

tor's own mind concerning it. In this Enderby insisted that he had done no better than choose the least of the evils presented, and that, somehow, some day, it behooved him to own the compromise made with his conscience. He did not see the way nor the hour, but he hoped that he was holding himself in readiness.

Early in the winter, the one vindictive foe of Royal Langbrith's memory perished in Mrs. Southfield, who had, indeed, only a conjectured, or, as she believed, an inspired grievance. Such as it was, she wished to visit it on the sinner's son rather than the sinner himself. Royal Langbrith had necessarily lapsed beyond her active hostility, and she turned this upon James Langbrith, whose engagement to Hope she never ceased to oppose. Hope herself took the humorous view of her grandmother's opposition, as she had taken the humorous view of her father's long tragedy, not because it was not real and terrible, but because temperamentally she had no other way of bearing it, because in that way she could transmute it into something fantastic, and smile at what otherwise must have broken her heart. She did not try to reconcile her grandmother to what her grandmother held her weak recreancy, but she reconciled herself to her grandmother, and assented and coaxed and had her way, and kept Langbrith from offering his antagonist a vain and exasperating propitiation. Mrs. Southfield's antagonism endured to the end. On her death-bed she left Hope a hoarsely whispered warning against the Langbrith tribe, as her last charge.

She might be said to have died of her vivid sense of a vague and unavenged injury, but her injury died with her, and with her died the sole reason against Hope's marriage.

There were people who contended for the fact of an unbecoming haste in her marriage, but these in their censure made no provision for the life of the girl, otherwise left absolutely alone in the world. Mrs. Enderby led the party against them, and with the support of Mrs. Garley, and their respective husbands, declared that Hope should not observe a vain decorum in waiting for a certain period of mourning to pass. She was married from the rectory, which Mrs. Enderby had made her make her home, three months after her father's death, and something less than three weeks after her grandmother's, and she went at Christmas to live with her husband in his father's house. Mrs. Enderby would have liked to infer a mystical significance from the coincidence of the event with the sacred time, when peace on earth and good-will was prophesied in every sort. If Dr. Enderby had been still a Unitarian, she would have openly done so, but under the circumstances she was not sure how far she might loose her imagination without compromising some doctrinal position of his, or committing him to what he might have felt a sentimental fancy. She confined herself to suggesting the notion to him, and contented herself with his assent that they might tacitly draw what comfort they could from the notion.

She did not feel it right to share it with Hope, but she permitted herself to share fully with the

girl the promise of her new happiness. There was
no question of primacy, in the house where Hope
went to live, between the elder and the younger Mrs.
Langbrith. People are modified rather than essentially changed, and it would be fatuous to pretend that James Langbrith was not irked in his
love of fitnesses by his wife's continuing in certain
things her relation of guest to the house where she
was really mistress. She left her mother-in-law
the head of the table, and the poor woman whose
life had always been in such an abeyance seemed
to satisfy an instinct of dominance, never gratified
before, in this shadowy superiority. The two worked equally together in other things of the house,
and there was no change except a turning, so grad-
ual as to be almost imperceptible, of the old Norah
and of Mary, the cook, to the younger Mrs. Langbrith for instructions.

The change did not awaken any apparent jeal-
ousy in the passive nature of the elder woman,
whose bearing towards her son betrayed no trace of
the past conflict of her weak will with his strong
will. At times, when he feared himself to have been
almost obviously impatient with her illusory head-
ship, or when Hope interpreted his restiveness to
him in that sense and blamed it, he sought little
occasions of reparation. But those seemed to afflict
her, and Hope had to warn him against being apparently other than he had always been to her. He
had to bear with that as he had to bear with another trial, which was less real. He had imagined
removing his father's portrait from its place over

the library mantel, but when he intimated his wish
to Hope she vehemently forbade it. That, she said,
was no more to be thought of, without the leading
that Dr. Enderby had insisted upon as Langbrith's
rule of action, than the removal of the commemora-
tive tablet from the front of the town library. They
must both stay till the providential time came.

As a matter of fact that time has never come.
The evil life of Royal Langbrith remains as he hid
it, except for the few contemporary and subsequent
witnesses of it. To the rest of the community
nothing is known; but as happens with men sometimes of whom nothing is known, there has grown
up in the public mind a certain conjecture of dis-
credit. This may have sprung from chance expres-
sions of Mrs. Southfield, in her theoretical distrust
of the whole Langbrith tribe; she could not always
be silent before people; but what is certain is that,
from the moment of the dedication of the votive
tablet by the son, the myth of the father suffered
a kind of discoloration, not to say obscuration.
Nobody could then say whether he was really the
saint and sage that he was reputed, and of what
nobody can say the contrary can be affirmed without contention, with even some honor to the shrewd
conjecture of those who affirm it.

The silence of Royal Langbrith's widow continued
as unbroken as that of Anther in his grave. It was
so inveterately the habit of her life that she never
betrayed herself to Hope, and what passed between
her and her son is as if it had never passed. The
whole incident of her proposed marriage with the

man who was so truly her friend is without trace in her actual relation to her son. It may be that the forces of her nature exhausted themselves in the struggle to accomplish her happiness, or it may be that her happiness was never essentially involved, and that she submitted to her fate without the suffering which Mrs. Enderby preferred to imagine of her. She never spoke of Anther, and whether she ever thought of him in the tender reverence which was his due Mrs. Enderby could not decide. Sometimes she was intolerably vexed with Mrs. Langbrith, sometimes she was resigned to the submission in which she saw the life of Mrs. Langbrith passing. That, when she came to think of it, was not without its dignity; and it was not what Anther himself, she realized, would have had changed into a futile rebellion. She realized, in her most vehement emotion, that there were women who had been long happily married, and who when widowed lived on in the same silence concerning the happiness they had lost as Mrs. Langbrith kept concerning the happiness she never knew.

Whether she duly enjoyed the happiness of her son in his wife was another question which vexed the kindly witness; but she saw that at least Mrs. Langbrith lived in harmony with them, and that a quiet pervaded the whole household which might very well pass for peace. After a certain period, which John Langbrith himself fixed for the instruction of his nephew in the business of the mills, James Langbrith took charge of them, and released his uncle to that voyage round the world in whose course he was to lose his dyspepsia, perhaps, with that equatorial day which lapses from the circumnavigator's calendar. He lost the day, if not the dyspepsia, and he returned with strength sufficiently renewed to bear it, which is probably the only real form of cure known to suffering. He then offered to let his nephew go back to Paris, if he wished, and resume his dramaturgical studies. There had been no explicit reconciliation between them, but a better reciprocal knowledge had done the effect of this, and it was with a respect for his nephew's ambition which he had not felt before that John Langbrith proposed to take up his job again in its entirety. The younger man did not respond directly. He asked his uncle, who had stopped in Paris on his way home, how Falk seemed to be getting on, and John Langbrith said Falk seemed to be doing well, and was at any rate working like a beaver; he had made a study of this fact, for he knew that James was paying his friend's way, and he did not want him to waste his money. He was not a judge of painting, but he was a judge of working, and Falk was working.

James Langbrith asked, "Did you have any talk with him about me?"

"Yes, I did," the uncle said, more promptly than willingly.

"What did he say?"

"Well, he said he would like to have you back, but—"

"Well?"

"If you really meant business, you could write

362

plays in Saxmills as well as in Paris. You could get it out of you, anywhere, if you had it in you."

James Langbrith did not ask if Falk had said anything of Susie Johns; he knew from Hope that their affair had been one of those without seriousness on either side, which pass with our young people in frequent succession, failing to eventuate in the matrimony which would be otherwise universal among us — without attaching blame to either side. There was something else that interested the young man infinitely more in the things that his uncle volunteered to tell him. John Langbrith, with greater reluctance than could have been predicated of him, either by himself or others, approached a fact which he said James ought to know, and when, without further preamble, he came out with it, his nephew agreed with him. One day, at his hotel in Paris, he had received the visit of a lady who seemed at first disposed to make a mystery of herself. She was the widow, she said, of a gentleman who had so far deceived her in marriage as not to have left her, at his death, so well provided for as she had expected, and she bore more heavily upon his want of candor in this respect than her own in another, though she was presently obliged to be entirely frank with John Langbrith. She was, it then appeared, the mother of that other family of his brother, who had provided for her so well that she was able to figure as a widow in easy circumstances when contracting her subsequent marriage. But her money had gone in the speculations which her husband was always engaging in for the increase of his fortune, and, if her children had not been nearly all provided for in successful marriages, she would not have known what to do. She did not know what to do now, in the case of the daughter whom she had with her in Paris for the cultivation of her voice. She had, as she said, kept track of Mr. Langbrith's family, and she had heard that he left a son—by another marriage, as she said; for in the retrospect she preferred to treat Royal Langbrith's relation to her as bigamous—very comfortably off. Without actually putting it, she left with John Langbrith the question whether this son might not like to do something for his sister, and, without actually putting it, John Langbrith now left the question with his nephew.

After a moment, James Langbrith asked, with a sickened face, "Did you see the girl?"

"Yes, I did."

"What sort of a girl was she?"

"About the sort her mother was, I guess, at her age. Why not?"

"Did you hear her sing?"

"She can sing all right, I guess. Maybe that'll keep her straight. Any rate, it don't seem to matter so much in that line of life."

"Yes," James Langbrith assented, from the dark, unwilling knowledge of the theatre which in the way of his ambition had revolted more than it had ever interested him. He added, "I will speak to Hope," and John Langbrith being apparently as sick of the subject as himself, they dropped it.

James Langbrith took it up again that night with

364

his wife, recurring to the general fact in his father's history with the shrinking which he did not understand her not understanding. When he had got the fact before her, "What ought I to do?" he asked, with a frown of disgust, as at some loathsome sight.

"You ought to tell your mother, in the first place," Hope said, and he answered, with still stronger repulsion:

"I don't see how I can."

"No," she assented. "I guess I shall have to do it for you," and Langbrith perhaps never felt so deeply her goodness and greatness as in this. With her wifely instinct, and the motherly instinct which was prophesying in her heart, she made known the fact to that virginal nature which never otherwise approached it. Mrs. Langbrith perhaps never fully realized the relation that established itself between her son and his father's past in his assumption of his father's cast responsibility, but Langbrith did so to the last fibre of his being. He needlessly stipulated with those people, as he always characterized them in his thought, that the recognition of the tie acknowledged was to be absolutely tacit; they had really no more wish to have it known than he; but at the bottom of Hope's heart there was what must be called a curiosity concerning her half-sister-in-law which she did not venture to own till she had Langbrith at disadvantage where he was helpless. It was when they hung together one night over the cradle of their first-born, and felt the holiness of her innocence purify their hearts, that she said, dreamily, " If she were the child of people who

365

had done wrong, I suppose she would be just as pure and sweet."

"What do you mean, Hope?" he cried, and she told him how she often thought of that girl, and how she longed to know what she was like, or what she looked like.

"Hope," he asked, "have you ever told Mrs. Enderby?"

"Indeed, I haven't!" she said, and then, woundedly, she asked, "Do you think I would speak of it without your knowing?"

"No, and I beg your pardon. I will ask the woman to send her picture."

But when the picture came, with the girl in the pose of the first part that had been given her in comic opera at Milan, which it had been her pride or her mother's to perpetuate in photography, Hope first gave the laugh that had so often defended her against the trials of life, and then prepared to break the blow to her husband.

He only glanced at the picture and said, without offering to take it from her, "We must keep on with the allowance," as if it had been in his mind instantly to withdraw it. He never asked her what she did with the picture, but when she had put it definitely away she remained with a longing to laugh herself over with somebody, in view of this oversatiation of her curiosity. She resisted her impulse to such a confidence with Mrs. Enderby not only because she was bound in honor against it, but because she did not believe Mrs. Enderby could enter perfectly into the spirit of the affair.

366

The wife of the rector, and through her the rector himself, continued in that patience with Providence which those more intimately concerned were obliged to practise in a situation of apparently indefinite duration. Enderby's patience was more tacit than that of Mrs. Enderby, with whom it often took the form of inquiry whether he thought there would ever be any revelation of the secrets of Royal Langbrith's life. She alleged that passage of scripture to which she had recurred from the beginning of her own privity. "There is nothing covered that shall not be revealed, and hid that shall not be known," and required him to reconcile it with the case in hand. Though she had agreed with Hope about that when the girl first offered her interpretation of the text, she had since had her recurrent misgivings, and she wished for a fuller exegesis from her husband.

Enderby was loath to put his wife off with the only answer he could make, and to say that, in the spiritual continuity of existence, eternity was not too far a term for the judgment of offences. He did not suffer with her at the hold which a bad man's life had kept after his death on those who survived him, and he reasoned in vain that good, evidently, and not evil, had come to others from leaving his life where the man himself had left it. In her soul she would have been willing the justice she longed for should have included the innocent as well as the guilty, but he gave her pause by making her reflect that in this instance earthly justice would include the innocent alone.

"Then you mean," she persisted, "that it must all go over to the day of judgment?"

"You know," he returned, "that I never like to say those positive things. But if we suppose that there is a day of judgment in the old sense, what else could it be for except for those sins on which justice has apparently been adjourned from the earthly tribunals?"

"There is something in that," she was forced to own.

"Besides, how do we know that upon this particular sinner justice has not already been done?"

"Why, what ever happened to *him?*"

"The fortitude of a man is no more the measure of his suffering than his weakness is. The strong suffer as much as the weak—only, they do not show it."

"Then you mean that Royal Langbrith suffered all that made his wife, and that other wretched woman, and Hope's father, and Dr. Anther, and poor James Langbrith suffer?"

"I don't say that. But could there be fearfuller suffering than his consciousness in his sudden death that he could not undo here the evil he had done? Why should we suppose him to have been without that anguish, if men in the presence of mortal peril are tormented with the instantaneous vision of their whole lives?"

Mrs. Enderby was silent, and measurably appeased. But then the rector went a step further, and in this it must be owned she could never follow him, great as her faith in him was.

"How do we know but that in that mystical legislation, as to whose application to our conduct we have to make our guesses and inferences, there may not be a law of limitations by which the debts overdue through time are the same as forgiven? No one was the poorer through their non-payment in Royal Langbrith's case; in every high sort each was the richer. It may be the complicity of all mortal being is such that the pain he inflicted was endured to his behoof, and that it has helped him atone for his sins as an acceptable offering in the sort of vicarious atonement which has always been in the world."

"But the blight—the misery he has left behind him?" she protested.

"Why," the rector said, "he seems to have left that around him rather than behind him. He made some of his own generation miserable — Hawberk and his wife, and his own wife, and that other woman, and Anther for them and with them. But Hope and James Langbrith are not unhappy. They are radiantly happy, and more wisely happy for tasting the sorrow which has not passed down to their generation."

"Then you don't believe that the children's teeth are set on edge by the sour grapes their fathers have eaten? What does the scripture say?"

"There are many scriptures, my dear. The scripture also says that the son who has not done the iniquities of the father shall not pay their penalty."

THE END

51